Handbook of Soil, Water and Plant Analysis

Handbook of Soil, Water and Plant Analysis

Dr. C. Seenappa
Associate Professor
Department of Agronomy
College of Agriculture
GKVK, Bengaluru-560 065

Dr. K.N. Kalyanamurthy
Professor and Head
Department of Agronomy
College of Agriculture
GKVK, Bengaluru-560 065

Dr. M.R. Anand
Assistant Professor
Department of Agronomy
College of Agriculture
GKVK, Bengaluru-560 065

Mr. Madam Vikramarjun
Sr. M.Sc. (Agri.)
Department of Agronomy
College of Agriculture
GKVK, Bengaluru-560 065

iBS

International Books & Periodical Supply Service

(Publisher of Scientific Books)
38, NISHANT KUNJ, PITAM PURA MAIN ROAD
DELHI-110034 (India)
Phone : 011-27352078, Mobile : 09810146811, E-mail : hkjain1975@yahoo.com

Published by :

International Books & Periodical Supply Service

(Publisher of Scientific Books)
38, NISHANT KUNJ, PITAM PURA MAIN ROAD
DELHI-110034 (India)
Phone : 011-27352078, Mobile : 09810146811, E-mail : hkjain1975@yahoo.com

ISBN 978-93-88892-07-0

Composed, Designed & Printed in India

University of Agricultural Sciences, Bengaluru

Gandhi Krishi Vigyan Kendra, Bangalore-560 065

Website : www.uasbangalore.edu.in
E-mail : vc@uasbangalore.edu.in
vcuasb1964@gmail.com
srprasad1989@yahoo.co.in

Off. : +91 080 2333 2442
Mob. : +91 94498 66900
Fax : +91 080 2333 0277

Dr. S. Rajendra Prasad
Vice Chancellor

Foreword

Depleting the natural resource base, deteriorating soil, air and water quality, declining crop yield and agricultural income triggered by imbalanced use of nutrients and increasing population have threatened the national food, nutritional, environment and livelihood security. Over the years, health of Indian soils has deteriorated resulting in decline in organic carbon content, soil biodiversity and soil physico-chemical properties and build-up of multi-nutrient deficiencies over the large area. It is reported that plant nutrient removal from soils by different crops is higher than addition from various sources, resulting negative nutrient balance. Soil nutrient deficiency and excess events are becoming the major production risk factors impacting agricultural production systems performance and management. This has necessitated the attention of scientists to develop outline to ensure sustainable productivity, profitability and efficient use of limited resources.

In today's overcrowded world, the challenge to feed the growing population of developing countries is a challenging task. Yields have to be increased from existing land areas; adding fertility to the soil to satisfy the demands of higher yielding crops is essential. Soils have inherent weakness, primarily deficiencies in nutrients that are essential to growing crops. Even when adequately supplied in the early stages of land cultivation, the nutrient supplying capacity invariably diminishes with time. Most soils are deficient in nitrogen (N); it is transient in nature, and plants need a lot of it. In many cases, phosphorus (P) is just as critical; soil chemical reactions reduce the effectiveness of P fertilizers. In recent years, there is a growing realization that other elements, e.g., micro-nutrients, are deficient in some areas of the region.

As no essential element can substitute for another, it is critically important to identify where and when such deficiencies occur. That's where the role of soil and plant analysis comes in. Techniques have been developed to evaluate soil fertility constraints based on soil chemical extraction and analysis of the plants that grow on such soils. Both are complementary and when calibrated with field crop responses to fertilizer, provide a rational basis to identify what elements are missing, and how much fertilizer, whether organic or inorganic, to apply. Therefore, soil and plant analysis laboratories have a vital role in global agricultural production.

We identified that the students and young scientists are quite dynamic but they find difficulties in the laboratories during analysis of soil and plant samples due to non-availability manuals and textbooks to know about the protocol and methodology for analysis. The present book titled "Handbook of Soil and Plant Analysis" suit to the need of students and teachers. I am happy that the authors have made efforts to write this handbook. In this connection, publication of book **"Handbook of Soil, Water and Plant Analysis"by Dr. C. Seenappa, Dr. K.N. Kalyana Murthy, Dr. M.R. Anand and Mr. Madam Vikramarjun** of UAS (B), GKVK, Bengaluru is quite appropriate and timely.

The idea for this soil, water and plant analysis handbook grew out and laid the basis for sound fertilizer recommendations. If soil, water and plant tests are not reliable, the process of sampling and analysis is meaningless and undermines the validity of any agronomic trials.

I congratulate the editors and the ICAR for bringing out this useful publication and hope that the students and teachers would benefit immensely from the book contents.

(S Rajendraprasad)
Vice Chancellor
University of Agricultural Sciences
Bengaluru-560065

Preface

"The nation that destroys its soil, destroys itself".

- Franklin Delano Roosevelt

The most valuable part of the mother earth is soil, which as a resource base for life on the planet. Thus, we have to maintain this important natural resource for our own survival. For this, the deficits generated on various accounts have to be made up. This is only possible when we know about the real situation prevailing at a point of time. Soil analysis come into picture here. Soil, plant and water are the major ingredients of agriculture system. Analysis of soil and plant as natural resources is of utmost importance for conservation strategies. To make things easy for students, we have attempted to bring all pertinent information together in this handbook. It has been compiled to give students thorough training the principles and methods of soil and plant analysis along with calculations and conversions of the values. The handbook is equipped with the estimation of mechanical and chemical analysis of soil and plant samples. The principles involved in various determinations have been explained and substantiated with modern concepts, giving chemical equations, wherever necessary. The handbook contains simple and reliable procedure which can be useful to students, teachers and scientists of agricultural sciences.

We hope the students, teachers and researchers in the field of agronomy and other related disciplines will find this handbook informative and useful, and we encourage all the readers to continue to provide feedback regarding its utility and errors, if any.

C. Seenappa
K.N. Kalyana Murthy
M.R. Anand
Madam Vikramarjun

Contents

Plant analysis for the estimation of

Chapter 1

Laboratory Safety Measures

The safety of the persons working in the laboratory and the accuracy of the experiments are very important during analysis. This can be achieved by following some guidelines/precautions :

1. Store the chemicals according to colour coding and regulations for storage and compatibility. Oxidizing and reducing agents should not be stored together. Do not store chemicals in alphabetical order. Acids should not be stored along with organic solutions. Read labels before opening a chemical container.

2. Develop a positive attitude towards laboratory safety and observe normal laboratory safety practices. Also, maintain a safe and clean work environment. Working table should be made dust free,every day morning wipe with a wet cloth.

3. While operating the instrument, follow the precautions provided by the manufactures. Monitor instruments while they are being used.

4. Avoid working alone. If one has to work alone, he should have someone to contact him periodically. One should learn what to do in case of emergencies-*learn emergency first aid :*

   a) If acid spills on the body of a worker, immediately wash with water.

   b) If there is short circuit, put off the electrical mains.

   c) If cut injuries occur, wash with alcohol.

5. One should seek medical attention immediately, if affected by chemical. Use first aid till medical aid is available.

6. Use forceps, tongs or heat resistant gloves to take out containers from hot ovens or muffle furnaces.
7. Do not eat, drink or smoke in the laboratory. Also, don't use laboratory glassware for eating or drinking. Do not store food in the laboratory.
8. Concentrated acids and other hazardous chemicals should be handled in fume covers.
9. Muffle furnaces and Atomic Absorption Spectrophotometer (AAS) must be vented to atmosphere.
10. Microwave oven should be checked for radiation leaks using suitable monitor.
11. Personal safety equipment's mentioned below should be used:
    a) Laboratory coat and chemical resistant aprons should be used.
    b) Hand glove to handle concentrated acids, bases and hazardous chemicals to protect hands.
    c) Dust mask should be used while grinding soil and plant samples.
    d) Always wear shoes (sandals should not be worn) in the laboratory to protect legs.
12. While diluting, always add acid to water but not water to acid.
13. Cylinders of compressed gases should be secured at all times.
14. One should not open a centrifuge cover until the machine stops completely.
15. Don't pipette by mouth (especially of concentrated bases, acid and organic solvents) and use bulb pipettes or automatic dispensers.
16. Extreme care should be taken while using per chloric acid to avoid fire or explosions. Handle it in fume hoods. While digesting soil or plant samples pre-digest the sample with nitric acid to destroy easily oxidisable organic matter. Don't allow per chloric acid to come in contact with rubber foliage and filter paper.

Chapter 2

Quality Control and Assessment Procedures

1. Good laboratory practices, such as storage of chemicals, laboratory techniques and good management practices, such as calibration, maintenance of equipment are integral parts of quality control. All chemicals are to be dated on receipt and disposed off upon expiry date.
2. Working standards should be prepared regularly.
3. Glass and plastic wares should be washed with dilute HCl or chromic acid, then with water and rinsed with distilled water and arrange the glass wares in dust free place.
4. Take all precautions in collecting proper representative samples. Sample integrity is most important (*sample should be a true representative of the population*).
5. Calibration of equipment is another important parameter.
   a. The pH meters have to be calibrated with two buffers.
   b. AAS is calibrated with standards for every batch of samples.
6. All details of analytical work (work sheet) should be filled as permanent records.
7. Number of significant integers: only the last figure reported should be in doubt.
8. Use appropriate blanks to check for contamination of filter paper, acids, water and blank sample should be run after every 20-50 samples.

9. Have duplicate samples to check for reproducibility of the results (duplicate samples for every 20 samples).

10. To ensure valid data, know reference materials or internal reference materials (samples collected, prepared and analyzed by several analysts) are run with each batch of samples. Adopt corrective measures, if results are not acceptable.

Chapter 3

Laboratory Glassware and Tools

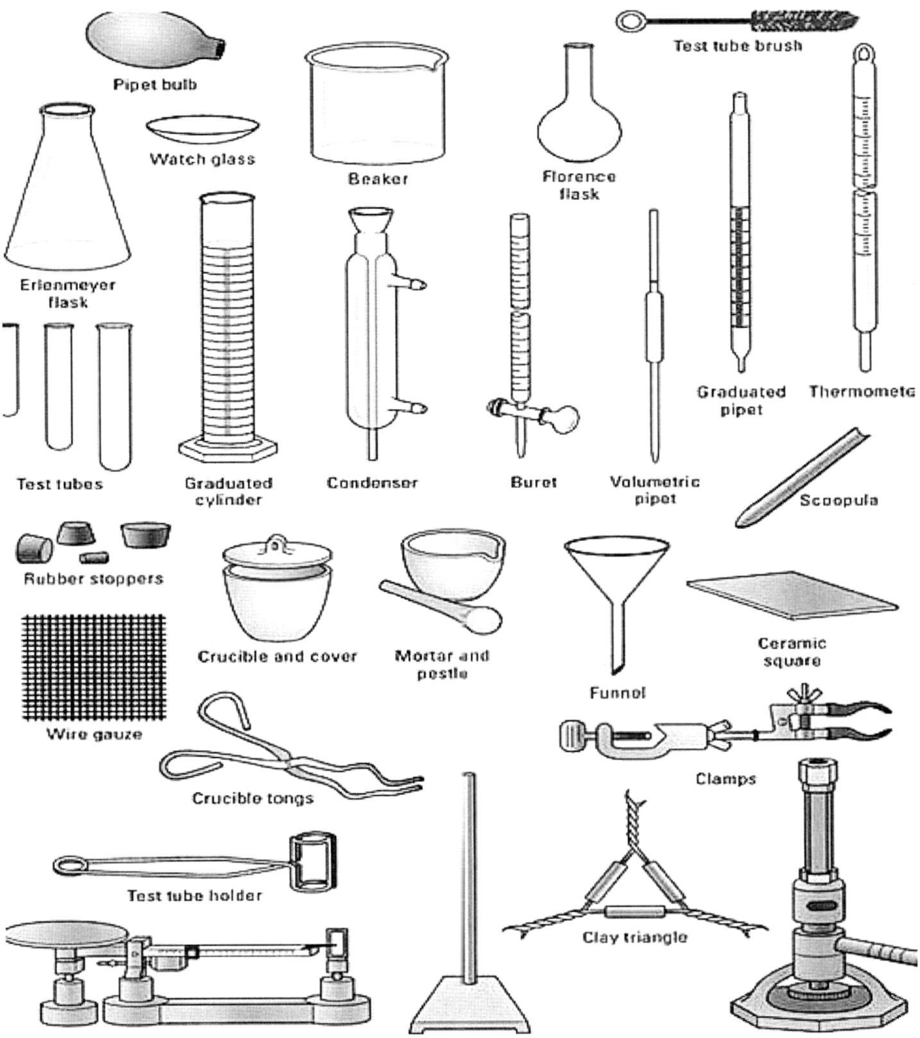

Platform balance (Triple beam) Ring stand Burner

Chapter 4

Soil Sampling

Soils are highly heterogeneous, hence soil has to be analyzed for various properties (Physical, chemical and biological) so that the amount of fertilizer or amendments can be recommended to increase crop production. Since, it is not possible to test the entire field, one must depend on samples taken from the field for testing.

Soil sampling is a technique by which a true representative sample of a given area is collected. Collection of representative sample is most important in an effective soil testing programme as the entire analysis and recommendation depends on the sample collected. Unless care is taken in collecting a representative sample, any amount of precaution taken during soil analysis will be a waste.

The type of sample to be collected depends on our objectives and the type of crops :

a) For study of soil genesis, soil sampling has to be done horizon wise from a profile

b) For studying soil fertility, sampling has to be done from the plough layer (0-6 cm)

c) For shallow rooted crops, samples has to be collected from the plough layer, but for perennial crops two samples at two depths(upto 30 cm). Depth of root penetration is an important consideration for deciding the depth of sampling. For field crops, a depth of 15-20 cm is desired, for deep-rooted crops like sugarcane, redgram, cotton and horticultural crops, it is necessary to have samples from different depths.

The depth of sampling should be decide based on the type of nutrient to be analyzed. For immobile elements like P, K^+, Ca^{2+} and

Mg^{2+}, sample upto 15 cm (tillage depth) is enough. But for mobile elements (NO^-_3 and SO^{-2}_4), some recent studies have indicated that samples should take upto a depth of 60 cm. In case of salt affected soils, visible salt crust on the soil surface should be sampled separately and the depth of sampling recorded.

The tools required for soil sampling are: Soil augers (post-hole and screw augers), soil tube, core sampler, spades, etc. Depending on the soil conditions and the amount of sample required, a suitable tool can be chosen. When the soil is moist, it is better to use a screw auger to get a good representative sample.

The steps involved in soil sampling are :

1. Divide the field into sampling areas so that each sample represents an area of not more than 6 acres. Each sample area should be uniform in soil type, slope, cropping pattern and past fertilizer management.
2. For sampling spots (about 15 or 20) at random in a zigzag pattern to represent the entire area. Avoid sampling near field bunds, wet places, trees, roads, drainage ditches, etc.
3. Scrape away the surface litter, stones and collect samples into a bucket from each spot upto the required depth by making a 'U' or 'V' shaped cut upto required depth using a spade and taking a slice of soil from both the sides. Collect same quantity of sample from each of the sampling spots.
4. Mix the sample on a plastic sheet discarding stones, roots, etc. and take out the required quantity of soil into a polyethylene bag by adapting quartering technique.
5. Fill in the information sheet giving the following details:

A. Area :	E. Survey No. :
B. Village :	F. Date of sampling :
C. Taluk :	G. Depth of sampling :
D. District :	H. Name of the farmer :

Sampling should be done after the harvest of crop so that there will be sufficient time for analysis sampling and making recommendations. The sample should be dried in shade at 20-30°C and with 20%-60% relative humidity ground using a pestle and mortar (taking care to break only the aggregates but not the sand and gravel particles) and passed through a 2 mm sieve to obtain the fine earth for further analysis.

Fig. 4.1: Soil sampling for analysis of nutrients

Note

1. Samples collected should be representative of the area sampled.
2. Adequate precautions to be taken while drying and preparing the soil samples to avoid contamination.
3. The pestle and mortar may be of wood, porcelain, agate or stainless steel.
4. The sieve may be of stainless steel, or nylon. Copper and brass sieves are to be avoided.

Chapter 5

Soil Sampling for Precision Farming System

Introduction

Soil sampling is the first step in generating field-specific information on which to base lime and fertilizer decisions and in monitoring soil nutrient status over time. For general soil sampling information, see the Soil Facts publication *Careful Soil Sampling: The Key to Reliable Soil Test Information (AG-439-30)*. Interpretations of soil tests are based on years of field calibration trials with crops grown on relatively small, uniform experimental plots. Since soils within large agricultural fields vary, composite soil test results cannot accurately represent all portions of a field. The composite field sample results mask scattered areas of both higher and lower levels of plant nutrients. If nutrient status is highly variable, then a substantial portion of the field area might respond to lime or fertilizer applications or both, even if a composite sample suggests no response.

Technological advances in GPS (Global Positioning Systems) and GIS (Geographic Information Systems) let producers design more intensive soil sampling schemes and to use this information for lime and fertilizer management decisions. Precision soil sampling provides information to:

- Identify localized regions of nutrient deficiencies and excesses within field
- Increase lime and fertilizer use efficiency by direct application to specific sites and
- Increase nitrogen use efficiency (which reduces leaching and runoff losses) by correcting other nutrient deficiencies

In contrast, the analysis of a composite soil sample from a field provides lime and fertilizer response data for average field conditions. Since fields are not uniform with respect to acidity and nutrient levels, each field consists of regions with differing lime and fertilizer needs.

Precision soil sampling terms

Base Map: Geo-referenced boundary map used repeatedly for location referencing.

Cell sample: Sampling throughout a grid cell so that the sample represents the entire cell.

Differential correction: Correction of GPS signals to increase accuracy from approximately 100 yards to 1 yard. This is done with an additional signal from a dedicated transmission tower (Mobile-Data, SatLoc) or a commercial FM radio station (DCI, ACCQPOINT) both of which have ranges of 100 miles or less, or by a separate corporate satellite signal (Chance, SatLoc, Deere), which has a range of thousands of miles.

Directed sample: Sampling scheme based on field variability due to either inherent soil properties (soil texture, drainage) or land management history (drainage, land shaping, spreader patterns, previous land-use). This reduces the number of samples while still recognizing zones of differing nutrient status.

GIS: Computerized system to store, manipulate and display geographically referenced information.

GPS: References ground position based on satellite signals. There are 24 US department of defense satellites, which beam directional signals. At least four are in view at anyone time. Intentional errors introduced into the signals limit accuracy to approximately 100 yards without differential correction.

Grid: A network of lines superimposed on a field map to assist in sample positioning.

Lat/Lon: Latitude (0-90° north or south of the equator) and Longitude (0-180° East or West of Greenwich, England). Can be given in degrees, minutes, seconds; degrees + decimal places; or a standard planar projection. In North Carolina, 1 longitudinal degree = 60 miles, 1 longitudinal minute = 5,280 feet, and 1 longitudinal second = 88 feet. Note that in North Dakota, 1 longitudal second equals only 72 feet, since it is closer to the North Pole, where the longitudinal lines converge.

1 latitudinal degree = 70 miles, 1 latitudinal minute = 6,160 feet, and 1 latitudinal second = 103 feet.

Point sample: Highly localized sampling around grid line intersections or around a point within a grid cell, so that the sample represents the point.

How to collect geo-referenced soil samples

1. Construct a base map of the field that includes geo-referenced boundaries

To construct a base map that delineates the field boundaries in a digitized format, buy the portable differential correction GPS navigation units that identify horizontal position to within 10 feet or less. Map any recognizable field subunit boundaries, if they are to be used in the design of soil sampling schemes. Relevant field subunits could include soil survey map units, or areas with distinct management history, consistently different crop yields Figure 5.1(a). Notice that Figure 5.1(b) reveals that a single grid system defines grid cells composed of only soil A, only soil B, and a mixture of soils A and B. This scheme may adequately reflect the nature of the transition zone between soil types. Or, each subunit could be sampled with a separate grid system preventing any overlap if desired, as shown in Figure 5.1(c).

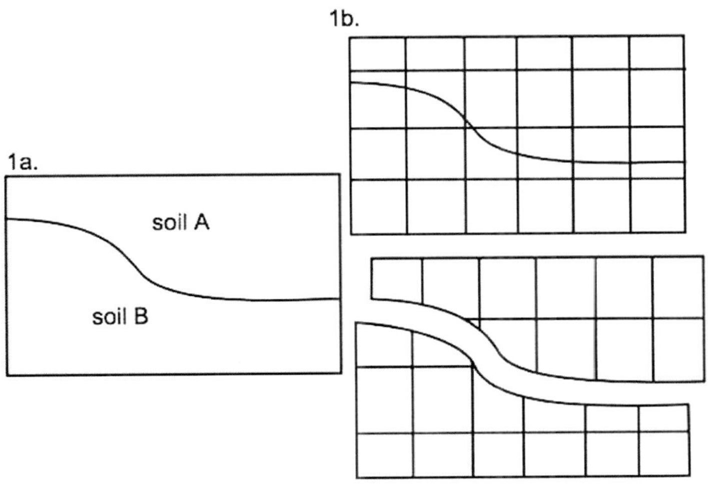

Figure 5.1: Base map showing (a) whole-field and subunits, (b) single overlaid sampling grid and (c) separate overlaid sampling grids for each soil type.

2. Select the best sampling strategy: cell, point, or directed sampling

Cell sampling : The most straight-forward sampling scheme is the "cell." A cell is a subunit of a whole field. Soil cores collected from locations throughout a cell are mixed to generate a composite sample for the cell. The resulting lime and fertilizer rates will be applicable to this entire cell. The entire field might then be represented by a checkerboard pattern of different recommendation rates. Currently, a grid size of 2 to 2.50 acres is commonly used for grain fields by North Carolina firms offering precision soil sampling services. This cell size is a practical decision, there are no mathematical criteria governing minimum cell size.

Smaller cell sizes might be justifiable, if a higher value horticultural crop is to be produced. In a 2.50 acre cell, 5 to 10 cores should be collected from random locations within each cell Figure 5.2(a). Or, they may be taken from a pattern which facilitates vehicular access to the sampling points, as shown in Figure 5.2(b). To avoid sampling position bias, do not concentrate sampling in the vicinity of a central point or sample within a straight line Figure 5.2(c) and Figure 5.2(d). Samples taken within a straight line might unintentionally correspond to spreader patterns of previous lime or fertilizer applications, or both.

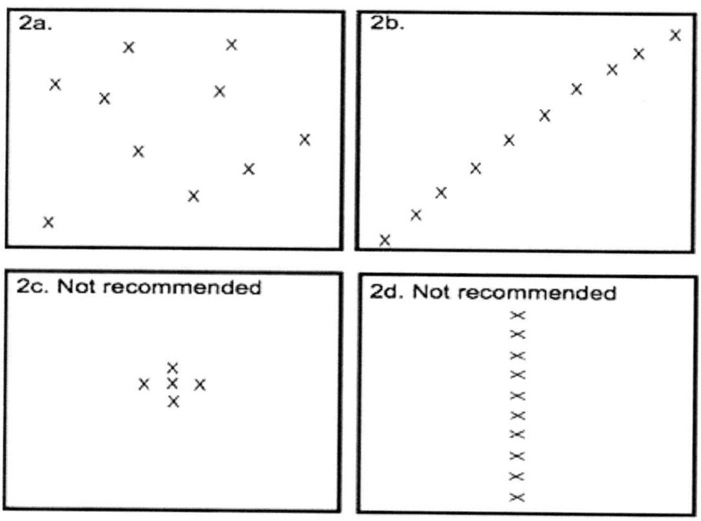

Figure 5.2: Base cell sampling patterns on (a) random core samples or (b) systematically arranged core samples to facilitate sampling vehicle access. Avoid (c) highly concentrated core samples and (d) core samples parallel with crop rows.

Point sampling is better for detecting patterns of field variability because all core samples are collected near geo-referenced points (located at grid line intersections), rather than scattered throughout the cell. Soil test parameters are calculated between sampling points, which permits the construction of contour maps of each soil test parameter (Figure 5.3). A disadvantage of the point sampling scheme is the close sample point spacing required for the mathematics to be reliable. Experience in the Mid-western US and from North Carolina suggests that points should be spaced 100 to 200 feet apart. At a spacing of 209 feet, one sample would be required for each acre of land, and at a spacing of 104 feet, four samples would be required for each acre of land. This is much more intensive, and thus, much more costly, than the more commonly used 2-2.50 acre cell sampling scheme for grain crops.

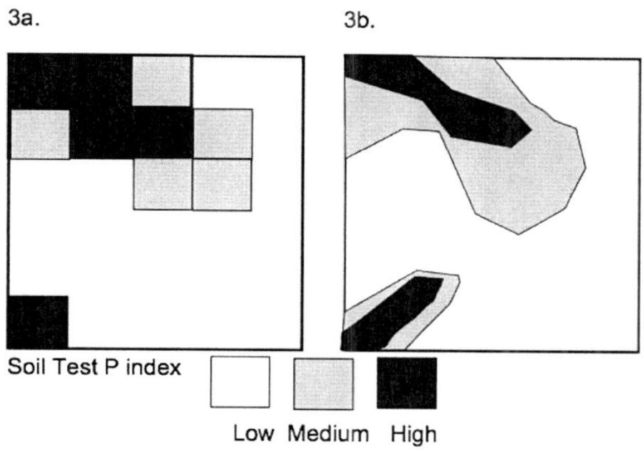

Figure 5.3: Soil P index maps (a) based on a cell sampling scheme and (b) based on a point sampling scheme.

Directed sampling uses an understanding of field variability to reduce the number of samples while still recognizing zones of differing nutrient status. Crop growth and yields vary due to a number of factors. Some of these are inherent soil properties (soil texture, drainage, etc.), and some are due to land management history (drainage, land shaping, spreader patterns, previous land-use, etc.). If a consistent soil factor controls yield variability in a field, then the distribution of this factor, or the distribution of crop yield, can assist in the design of a soil sampling scheme (Figure 5.4). On a cautionary note, if the selected factor is not a consistent predictor of yield, this scheme unnecessarily biases the sampling process and is less efficient than either cell or point sampling schemes.

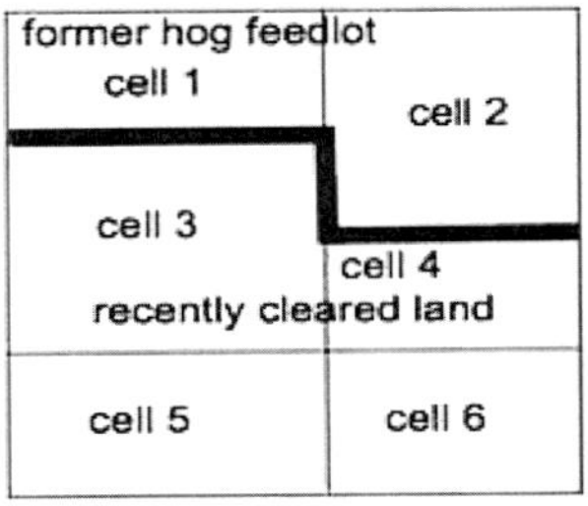

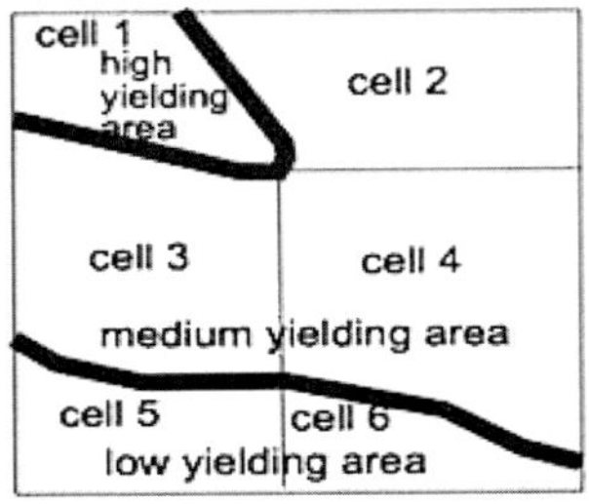

Figure 5.4: Directed soil sampling schemes (a) based on previous land-use and (b) based on prior crop yields.

There are no definitive criteria dictating, which soil sampling scheme must be used. The following table 5.1 describes some of the advantages and disadvantages of each of the three sampling schemes.

Table 5.1 : Precision farming soil sampling schemes

Sampling Scheme	Advantages	Disadvantages
Cell	Fewer sampling biases No cell size restrictions	Poor characterization of variability, i.e. coarse checkerboard pattern results
Point	Better characterization of field variability	Requires closely-spaced samples
Directed	Variability characterized with reduced number of samples factor	May be biased if scheme is not based on yield-determining

2. Generate sampling grid with appropriate shape, size, and orientation

Computer software packages can position sampling grids on top of the field base map. A special consideration of the long (2500-plus feet), narrow (150-300 feet) fields in the North Carolina Tidewater region is that rectangular grids are more appropriate than square grids (Figure 5.5). This is due to the management induced variability along the narrow axis, when digging surface ditches and crowning fields. Thus, variation occurs over a shorter distance along the narrow axis than along the long field axis. For point samples on these tidewater fields, spacing of 100 feet are required in the narrow dimension, while spacing of 700 feet along the long axis are sufficient. In all fields, grid sizes and compass direction, orientations can be manipulated to more closely align grids with field borders.

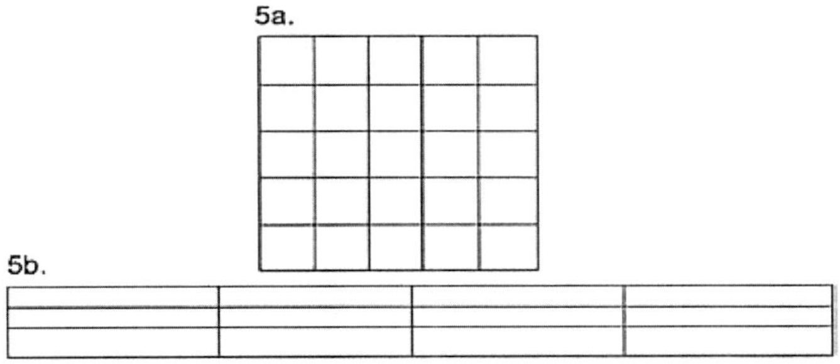

Figure 5.5: Sampling grid shapes for (a) common fields and (b) long, narrow Tidewater fields with surface drainage ditches.

3. Collect soil samples using appropriate procedures

As with all soil sampling, attention to sample depth and collection of sufficient cores are critical. The North Carolina Department of Agriculture and Consumer Services recommends that samples from cultivated fields be collected to the depth of the plow layer, generally 6 or 8 inches. For established no-till fields and sod, samples should be collected to a depth of 4 inches. Traditional composite soil samples of 20 acres or less should consist of 15 to 20 cores, or approximately 1 core per acre. With a precision soil sampling scheme based on 2.50 acre grids, 5 to 10 cores should be collected, or approximately 1 core per $^{1}D_{4}$ to $^{1}D_{2}$ acre. Avoid fertilizer bands and areas disturbed by vehicular traffic. Thoroughly mix the core samples in a clean plastic bucket and then transfer the sample to a soil box.

Costs and benefits

An informal survey of providers of precision soil sampling services in Eastern North Carolina revealed a cost range of $8.00 to $9.50 per acre for sample collection on a 2 to 2.50 acre grid cell size. One provider offered 1acre grid cell sampling for $19.50 per acre for vegetable producers. The cost for traditional, composite soil sampling varies with field size, but general estimates are $1.00 to $2.00 per acre. Cost estimates for variable rate applications of lime and fertilizer are difficult to standardize, as they can be given in terms of $/ton or $/acre spread. Acres spread are less than total field acres, if not all of the land receives lime or fertilizer. General estimates of the additional cost for variable rate application above those for conventional application in North Carolina are approximately $2.00 to $3.00 per ton for lime, and $3.00 to $5.00 per ton or $1.50 to $2.75 per acre spread for fertilizer.

Several potential benefits of precision soil sampling and variable rate applications exist. Variable application rate technology should minimize nutrient deficiencies and reduce costs by eliminating or reducing application rates at sites with sufficient lime or fertilizer, or both. The correction of localized nutrient deficiencies should lead to an increase in nitrogen use efficiency, since other nutrients will not be limiting. If these benefits can be met, then more efficient nutrient utilization should reduce runoff and leaching risks.

A key factor in realizing the potential benefits of precision soil sampling lies in improving our understanding of yield-limiting factors within fields. Yields can vary within fields for many reasons: weeds, insects, microclimate, soil nutrient status, and other soil properties (texture, topography, wetness, etc.). The greater the role of soil nutrient status in determining yield predictability, the greater the potential of precision soil sampling. Precision soil sampling is one of an array of precision farming tools, available to use nutrients more efficiently in agriculture.

Chapter 6

Mechanical /Physical Analysis of Soil

The process of determining the amount of sand, silt and clay in a soil is called mechanical analysis. It is one of the most important laboratory determination is made on soil as the knowledge of mechanical analysis gives an insight into the physical, chemical and biological properties of soils and also, helps in proper management of soil.

Several methods have been developed for mechanical analysis and all the methods are based on Stoke's Law with principle of differential rate of falling or settling of particles in liquid medium :

a) The international pipette method (most accurate and widely followed)

b) The hydrometer method (very rapid and less tedious)

c) The beaker method (time consuming and tedious)

Two essential steps involved in any of the method are :

a) Dispersion of the soil separates

b) Grading of dispersed particles into different sized groups.

Complete dispersion of soil is obtained by removing cementing substances responsible for aggregation. The respective cementing substances are removed by treating with H_2O_2 to oxidize the organic matter, $CaCO_3$ is removed by treating with diluted HCl. (finally make it free of chlorine by washing with distilled water) and iron oxides by sodium dithionate sodium hexametaphosphate (SHMP) and sodium carbonate (SC) solution as a dispersing agent is added before mechanical stirring with stirrer.

Reagents

1. Hydrogen peroxide (6%)
2. Dispersing agent (3.6% sodium hexametaphosphate + 0.8% sodium carbonate)
3. Dil.HCl (0.1 N)
4. Sodium dithionate.

Materials required: Measuring cylinder (1000 ml, stopper type), tall beaker (60 ml), glass rod, stirrer (0.2 mm), porcelain dishes and pipettes.

International pipette method

Procedure

1. Take 20g of air dry soil into a tall beaker, add 50 ml of H_2O_2 (6%), cover it with a watch glass and keep it overnight. Then, place it over a water bath at 60-65°C. Stir occasionally as the reaction subsides, add another 30 ml of H_2O_2 to ensure complete oxidation of organic matter.
2. Wash the material on the sieve with a jet of water till all the soil particles are washed into a beaker. Collect the materials left on the sieve (coarse sand) into a previously weighed crucible, dry and weigh (CS). To the beaker containing soil-water suspension add 10 ml of dispersing agent and stir the content of the beaker using a mechanical stirrer for 5 minutes and transfer the suspension to a 1000 ml cylinder through a 0.2 mm sieve. Make up the volume to 1 litre. Note the temperature of the suspension and from the table 3.1 given below, find out the time required for fine sand and silt particles to cross 10cm depth (sedimentation time).
3. Shake the contents and keep it undisturbed. After the fine sand sedimentation time. Low the pipette to a depth of 10cm from the surface and draw out 20ml of liquid slowly (in 20 seconds) into a previously weighed (W_1) dish and dry the content on a water bath, cool and weigh (W_2). Calculate the silt + clay per cent (A), as the sample contains both silt and clay suspensions.
4. Shake the suspension again and keep it undisturbed for silt sedimentation time. Collect the sample (20ml) has done in step. 4 and transfer it into a previously weighed dish (W_3) and weigh (W_4) after drying.

5. Decant or siphon out a major portion of the suspension and transfer the remains into a tall beaker using a jet of water. Make up the volume to a height of 10cm. Decant the clear supernatant and transfer the residue to previously weighed crucible (W_5), dry the contents and weigh (W_6) from these values, calculate the per cent of sand, silt and clay.

Per cent coarse sand = $\frac{CS}{20} \times 100$

Per cent silt + clay (A) = $\frac{(W_2 - W_1) \times 1000}{20 \times 20} \times 100$

Per cent clay (B) = $\frac{(W_4 - W_3) \times 1000}{20 \times 20} \times 100$

Per cent silt= A-B

Per cent fine sand= $\frac{(W_6 - W_5) \times 1000}{20 \times 20} \times 100$

The textural classes are based on the per cent sand, silt and clay contents. The texture of the soil can be found using a triangular textural diagram.

Hydrometer method

Principle

The specific gravity of the soil suspension is proportional to the weight of soil particles in suspension and different soil particles will settle out of suspension at the rate proportional to their size.

Hydrometer method is the simplest and commonly used. Hydrometer consists of bulb and stem made of glass with a solid loading material cemented to its bottom part. The stem is graduated in g/1000cc suspension to get the percentages of soil remaining in suspension.

Materials required

Bouyoucos soil hydrometer (graduated in gL^{-1}), measuring cylinder, mechanical stirrer, dispersing agent, hydrogen peroxide (H_2O_2).

Procedure

1. Take 50g of soil sample into a 600 ml beaker.
2. Add 100 ml hydrogen peroxide and keep it overnight and boil the contents on a water bath at 60°C. Add another 50ml of hydrogen peroxide to destroy all soil organic matter.

3. Add 10ml of dispersing agent and stir the contents of the beaker.
4. Transfer the contents to a 1000ml measuring cylinder through a 0.2mm sieve. Wash the contents of the sieve with a jet of water to wash down all the fine sand, silt and clay content on the sieve.
5. Transfer the contents of the sieve to a previously weighed dish and determine the coarse and per cent, as mentioned earlier.
6. Make up the volume of the suspension to 1000ml and note the temperature. Shake the contents and keep it undisturbed.
7. Record the hydrometer reading for silt+clay and clay by dropping the hydrometer very slowly into the suspension after their respective sedimentation times. The hydrometer reading is noted only after its satisfactory or no motion.

Table 6.1: Sedimentation time for fine sand and silt fractions (For 10 cm depth and time in minutes, seconds)

Temperature (^{0}C)	Fine sand	Silt
8	6.40	11.00
9	6.30	10.40
10	6.20	10.25
11	6.10	10.10
12	6.00	9.50
13	5.50	9.35
14	5.40	9.20
15	5.30	9.05
16	5.20	8.50
17	5.10	8.35
18	5.00	8.25
19	5.00	8.10
20	4.48	8.00
21	4.40	7.50
22	4.30	7.25
23	4.30	7.25
24	4.20	7.15
25	4.15	7.05
26	4.10	6.55
27	4.05	6.45
28	4.00	6.40
29	3.55	6.30
30	3.50	6.20
31	3.45	6.15
32	3.40	6.05
33	3.35	5.55

The data given in table 54.1 (in Annexure) gives the depth of sampling for silt and clay (0.02 mm) at 5 minutes and for clay (0.002 mm) at 3 hours 30 minutes. This data is very much useful and all the analysts can follow this, as they can complete taking samples/ readings of silt+clay within one day.

Chapter 7

Estimation of Bulk Density and Particle Density of Soil

Density is defined as mass per unit volume of a substance. The approximate densities (g/cc) of some substances are: water-1; wood-0.7; quartz-2.6; steel-7.7; mud-11.3 and mercury-13.5. Water is the reference material for density measurements. For soils two densities are commonly used, and they are the bulk or apparent density (BD) and the particle density or true density (PD).

I. Bulk Density (BD)

The BD is defined as the ratio of the weight of a unit volume of oven dry soil to the weight of the same volume of water or it is the oven dry weight of a unit volume of soil inclusive of pore spaces and is expressed as g/cc or lbs/ft^3. The BD of soil indicate the degree of compactness and aeration status. It also helps to find out the weight of soil in acre or hectare furrow slice (AFS or haFS) and to convert soil moisture content on weight basis to volume basis.

The BD of soils generally varies from 1 to 1.8 g/cc. The BD of soils generally increases with depth due to the low organic matter content of the lower layers and due to compaction from the pressure of the upper layers and because of the use of implements and machinery.

Methods of determination

The BD of both disturbed and undisturbed soils can be determined and the latter one is more realistic. The determination based on the disturbed soil sample gives only an approximate value, which is of little significance and have limited application.

BD of disturbed soils (glass vial/weighing bottle method):

Principle

The mass of the soil is determined by weighing the oven dried soil sample. The soil is put into the container in small amount and gently tapped for several times. The tapping is assumed to produce the same intensity of packing as occurring naturally in the field. The volume of the packed soil will be equal to the capacity of the container. The BD is calculated from the data of weight and volume.

Apparatus required

Weighing bottle, analytical balance and burette

Procedure

Weigh an empty bottle (W_1 =bottle) of about 50cc capacity without the stopper. Fill the bottle with soil, as explained above. Weigh the bottle again (W_2=bottle + soil). Empty the bottle and add water to it from the burette and note down the volume of the water required to fill the bottle to the brim (V cc).

Observations and calculations

Weight of the empty bottle	=	W_1 g
Weight of the empty bottle + soil	=	W_2g
Weight of the soil	=	$(W_2 - W_1)$ g
Volume of water filling the bottle	=	V cc
Bulk density (g/cc)	=	$(W_2 - W_1)$ g/ V

BD of undisturbed soil

The BD is measurement by taking an undisturbed block of the soil clod (clod method) or soil core (core sampler method). The soil is dried and weighed and the volume is also determined and the BD can be calculated by the ratio of the weight to the volume of soil. In the clod method, the volume of clod is measured by water displacement method after coating the clod with paraffin or liquid plastic. In core sampler method, the cores are collected by metallic cylinder and the volume of the core sampler is calculated by cylinder volume ($V = \pi r^2 h$).

II. Particle Density (PD)

The PD of soil is defined as the mass per unit volume of the soil solids. It is the weight (g) of the solid phase and expressed in g/cc. The PD value of most of the mineral soils is between 2.6-2.7 g/cc with an average value of 2.65 g/cc. The presence of large amount of organic matter and heavy minerals will affect the PD. A knowledge of PD is important, in the volume-weight relationships of the soil like porosity, aeration porosity and void ratio. It is also important in calculating the settling velocity of particles (Stoke's law) of different sizes during mechanical analysis and are also useful in erosion research.

Methods of determination

The PD of soil determined by the ratio of the weight of the soil to the volume of the soil solids, the weight of the solids is obtained by weighing oven dry sample and the volume is by immersion (water displacement by each particle). The pycnometer method is widely used, in addition Priharand Sandhu (1968) developed a simple and rapid method.

Pycnometer method

When a dry soil weighing (W_s) is immersed in a definite volume of water and air is expelled out, it displaces volume of water equal to its own. The volume of particles is calculated from the weight of water (W_W), displaced by the immersed particles. For this, weight of water filled pycnometer (W_{pw}), weight of dry soil (W_s) and weight of pycnometer containing soil and weight of pycnometer containing soil and filled with water (W_{psw}) are recorded. Mathematically,

$$(W_w) = (W_{pw}) + (W_s) - (W_{psw}) = V_s \times D_w$$

Where, $V_s = W_s/D_p$; Hence, $D_p = W_s \times D_w/W_w$

Assuming the density of water to be 1g/cc,

$$\text{The PD} = W_s / W_{pw} + W_s - W_{psw}$$

Apparatus

Pycnometer, pipette (20 ml), analytical balance, water bath and filter paper.

Procedure

1. Take a clean, dry pycnometer, fill it with boiled, cooled distilled water, note down the temperature. Replace the stopper, and dry the outside of the pycnometer with piece of filter paper and weigh it.
2. Empty it and put into it a known weight of oven dried soil (10g) in the absence of oven dry soil, determine the moisture content of the soil separately.
3. Using the pipette fill the bottle to about half with water, washing into it, the soil particles sticking to the inside of the neck.
4. Remove the entrapped air by gently boiling the contents, in case of heavy clayey soil, leave the soil overnight so that soil pores become saturated.
5. Cool the contents to room temperature and fill the pycnometer with boiled distilled water cooled to room temperature.
6. Insert the stopper carefully, wipe the outside of the pycnometer with a filter paper and weigh.

Observations and calculations

Weight of water filled pycnometer = W_{pw} g

Weight of soil = 10g

Weight of pycnometer = W_{psw} g

PD = $10/W_{pw} + 10 - W_{psw}$

Chapter 8

Estimation of Maximum Water Holding Capacity and Porosity of Soil

Maximum WHC is the amount of water in a soil, when all its both macro and micro or capillary are completely filled with water. This acts as a rough measure of the total porespace of the soil.

Volume expansion of soil is defined as the increase in volume in cc per 100 cc of soil, when the soil is at its MWHC. This is influenced by the kind and amount of clay present in the soil (and has bearing on the soil structure).

Keen and Raczkowski (1921) modified the Hilgard's experiment and gave a single method for the estimation of BD, PD, pore space and volume expansion. The simple device used in the experiment consists of a circular brass cup of about 5cm diameter; 1. 5cm height with perforated base and a screw type ring just to fit in the cup as tightly as possible.

Materials required

Keen's cup, spatula, moisture can, tray, weight box, oven, analytical balance, filter paper, camel hair brush and watch glass.

Procedure

Measure the inner diameter and the height of the brass cup, fit in a filter paper and find out the weight of the cup along with a watch glass. Keep it on a sheet of paper, and pack the soil sample, putting small quantities at a time and with gentle tapping on the working table. Continue filling and tapping till the cup is full. Then remove the surplus soil with a sharp blade. Gently tap and add some more soil, if required and level it. Remove the soil from the outer surface with a camel hair brush and find out the weight by keeping it on a watch glass.

Slightly screw up the ring of the cup and place it with soil in a petridish or tray containing water to a ¼ and allow the soil to absorb water till it is saturated. Then take out the cup from the tray, wipe the outer surface with a dry cloth immediately, find out the weight by keeping it on the watch glass.

Next screw down the ring to its original position, scrape out the wet expanded soil by means of a clean spatula transfer to a previously weighed porcelain dish or watch glass and find out its weight by keeping it on a watch glass. Also, find out the weight of the cup with the remaining wet sample. Now dry in an oven (105°C) the soil present both in the cup and the porcelain basin and find out the weight.

Observations and calculations

Inner diameter of the cup	= d cm
Inner height of the cup	= h cm
Volume of the cup (v)	= $d^2h/4$
Weights of	
Cup+ filter paper (FT) +watch glass (WG)	= 'a' g
Cup + FP +WG + Dry soil	= 'b' g
Cup + FP+ WG+ Saturated soil	= 'c' g
Cup + FP +WG +Wet residual soil	= 'd' g
Cup +FP +WG +Residual soil after drying	= 'e' g
Porcelain dish	= 'f' g
Dish + wet expanded soil	= 'g' g
Dish + expanded soil after drying	= 'h' g

1. BD of soil = Wt. of oven dry soil/ Bulk volume of the soil
   (e-a) + (h-f)/V

2. PD of soil = Wt. of oven dry soil/volume of the soil solids

$$= \frac{(e-a) + (h-f)}{V - \{(d-e) + (g-h)\}}$$

3. MWHC = (Wt. of saturated soil – Wt. of oven dry soil) x 100

$$= \frac{\{(d-e) + (g-h)\}}{\{(e-a) + (h-f)\}} \times 100$$

4. Pore space $= \dfrac{\text{(Total weight of water in soil)}}{\text{(Bulk volume of soil)}} \times 100 = \dfrac{\{(d\text{-}e) + (g\text{-}h)\}}{V} \times 100$

5. Volume of expansion of the soil $= \dfrac{\text{(Volume of expanded soil)}}{\text{(Bulk volume of soil)}} \times 100$

$$= \frac{\{(g\text{-}h) + (h\text{-}f)/BD)}{V} \times 100$$

Chapter 9

Estimation of Organic Matter Content in Soil

The soil organic matter (SOM) is defined as the organic fraction of soil including plant, animal and microbial residues, fresh and at all stages of decomposition and the relatively resistant soil humus (SSSA, 1979). Now, SOM includes only those organic materials that accompany soil particles through 2 mm sieve. Its content is an index of the soil productivity.

The SOM is a source and sink of nutrients (N, P and S); it has charge properties which make it a site of ion exchange (generally, 1 per cent OM increase contributes 2 meq); it has physical and chemical properties which facilitate aggregation with mineral particles, particularly clays, and intern modify soil physical structure and influence water regimes; and it is a source of energy for soil biota and thus, influences many biologically mediated processes if soil. Thus, SOM itself represents a set of attributes rather than an entity and is renewable resource.

Methods

The SOM content is estimated from the organic carbon (OC) determination by any of the following principles :

- The difference is estimated between the total C and inorganic-C.
- The total-C determination after complete destruction of inorganic-C.
- Reduction $Cr_2O_7^{-2}$ to Cr^{3+} by OC compounds and subsequent determination of unreduced $Cr_2O_7^{-2}$ by redox titration or by colorimetric method.

The above can be broadly classified as :

a. Dry combustion method

b. Wet digestion method

The C is the chief element present in SOM, comprising from 48%-58% of the total weight. Determine the SOC and convert them to SOM by multiplying SOC with 1.724 (van Bemmelen factor) with the assumption of 58 per cent of C in SOM. The conversion factors for SOC to SOM varies from 1.724 to 2.0 and it should be determined quantitatively for each soil, rather than its conversion to SOM by approximate correction factor.

Soil organic carbon (SOC) fraction consists of microbial cells, plant and animal residues at various stages of decomposition, humus and highly carbonized compounds, such as charcoal, graphite and coal. Most of organic carbon is present in the SOM form and inorganic C in carbonate minerals.

1. Dry combustion method

A known weight of the soil (previously treated with sulphuric acid to destroy $CaCO_3$) is ignited in silica tubes packed with copper oxide. The CO_2 liberated is absorbed in a weighed soda-lime tube. The increase in weight of the soda- lime tube represents the total CO_2 evolved from the oxidation of SOM.

In another method, loss on ignition, the organic matter is oxidized by heating at 375°C and estimated by loss on weight basis.

Apparatus required

Muffle furnace, silica crucibles, desiccators, balance

Procedure

1. Heat silica crucibles for 1 hour at 375°C, cool in open to about 150°C, cool in a desiccators for 30 minutes and weigh.
2. Weigh 5 g of oven dried soil sample (passed through 2 mm sieve) into crucible.
3. Place the crucible containing the sample in muffle furnace, heart slowly by increasing the temperature.

It is a gravimetric, time consuming and tedious method. But it is relatively more accurate. Another disadvantage is the OC and elemental C present as charcoal (non-humus form) cannot be distinguished.

2. Wet digestion method

The oxidation of SOM is carried out in wet condition by oxidizing conditions.The oxidation of SOM is made by strong oxidizing agents, such as $K_2Cr_2O_7$ in acidic condition, H_2O_2, *etc.*

The chromic acid oxidation method is also known as Walkley and Black's method (1934). This method is simple, rapid, widely used and large number of samples can be analyzed in a day, and is not that accurate. Rapid $K_2Cr_2O_7$ methods are subject to interference by oxidizable soil constituents, such as Cl^-, Fe^{+2} and MnO_2. However, the presence of $CaCO_3$ in the soil does not interfere with this method. In addition, this method distinguishes elemental carbon from SOM.

Diphenylamine was the first redox indicator used in dichromate method. The end point in this titration can be improved by NaF or H_3PO_4 before titration. Jackson recommended orthophenonthroline (Ferroin) as an indicator in $Cr_2O_7^{-2}$ titration as colour change occurs at higher redox potential compared to diphenylamine.

Principle

A known weight of the soil is treated with an excess volume of standard $K_2Cr_2O_7$ solution in the presence of concentrated H_2SO_4. The soil is digested by the heat of dilution of H_2SO_4 and organic carbon in the soil is, thus, oxidized to CO_2. The excess of $K_2Cr_2O_7$ unused in oxidation is titrated back against a standard solution of ferrous ammonium sulphate {FAS; $FeSO_4\ (NH_4)_2SO_4$} in the presence of NaF or H_3PO_4 and diphenylamine solution (indicator).

Reactions

A. The oxidation of SOC

$$\frac{\{K_2Cr_2O_7 + 4\ H_2SO_4 \rightarrow Cr_2(SO_4)_3 + 4\ H_2O + (O)\} \times 2\{C + 2\ (O) \rightarrow CO_2\}\ x\ 3}{2\ K_2Cr_2O_7 + 8\ H_2SO_4 \rightarrow 2\ K_2SO_4 + 2\ Cr_2(SO_4)_3 + 3\ C + 6\ (O) + 8\ H_2O + 3\ CO}$$

B. The titration of excess nascent oxygen by std. FAS solution

$$\{FeSo_4(NH_4)_2SO_4 . 6H_2O \rightarrow FeSO_4 + (NH_4)_2SO_4 + 6H_2O\} \times 2$$
$$FeSO_4 + H_2SO_4 + O \rightarrow Fe_2(SO_4)_3 + H_2O$$
$$2\ FeSO_4(NH_4)_2SO_4 . 6H_2O \rightarrow 2(NH_4)_2SO_4 + Fe_2(SO_4)_3 + 13H_2O + H_2SO_4 + O$$

C. The action of diphenylamine in dicator

$$C_6H_5\ NH\ C_6H_5 \xrightarrow[-H2O]{+O} C_6H_5\ NH\ C_6H_4 \xrightarrow[-H2O]{+O} C_6H_5\ NH\ C_6H_4$$

Reagents required

1. **1 N $K_2Cr_2O_7$ solution:** Dissolve 49.04 g of potassium dichromate in distilled water and make upto 1 litre.
2. **0.5 N FAS (Mohr's Salt):** Dissolve 196 g of FAS in distilled water, add 50 ml Conc. H_2SO_4 and make up the volume to 1 litre.
3. **Diphenyl amine indicator:** Dissolve 0.5 g of diphenylamine in a mixture of 100 ml on conc. H_2SO_4 and 20 ml of distilled water.
4. **Concentrated H_2SO_4**
5. **Sodium fluoride (NaF)**
6. **Ortho phosphoric acid (85%)**

Apparatus required

500 ml conical flask, pipettes, burettes and measuring cylinders (25 ml)

Procedure

1. Weigh 0.2-0.4 g of 0.2 mm sieved soil (powdered and then sieved) into a 500 ml conical flask.
2. Add 10 ml of 1 N $K_2Cr_2O_7$ solution and shake to mix.
3. Add 20 ml of conc. H_2SO_4 from the sides of the flask.
4. Keep the contents of the flask undisturbed for 30 minutes.
5. Add 3 g of NaF or 10 ml of H_3PO_4 and 100 ml of distilled water and shake vigorously.
6. Add 10 drops of diphenylamine indicator, which gives violet colour.

7. Titrate against 0.5 N FAS solution till the colour changes from violet to bright green and note down the volume of solution used.
8. Carry out blank titration in a similar manner without soil.

Observations and Calculations

Weight of sample	= S	g
Volume of FAS used in blank	= X	ml
Volume of FAS used to oxidize SOC	= Y	ml
Normality of FAS	= N	
Volume of 1 N $K_2Cr_2O_7$ used for the Oxidation of Carbon	= (X-Y) /2	
1 ml of 1 N $K_2Cr_2O_7$	= 0.003 g of organic carbon	

Per cent of organic carbon in the soil (A) $= \frac{(X-Y) \times 0.003 \times 100}{2 \times S}$

Per cent of organic matter = A x 1.724

Additional information

The uses of heat of dilution don't give complete oxidation of organic compounds in soil. Walkely and Black (1934) observed about 60-80 % of the SOC was oxidized by the heat of dilution and the Correction Factor (CF) has to be developed to account for unrecovered org-C. Thus, the above method should be considered to give approximate or semi-quantitative estimates of SOC. In general, 1.31 CF is found to be most reasonable for most soils.

The presence of higher amounts of Fe^{2+} or Cl^- will lead to an over estimation as Cl^- utilizes chromate forming chromyl chloride and as Fe^{2+} gets oxidized to Fe^{3+}. Chloride interference may be eliminated by washing the soil free of Cl^- before analysis or by precipitation with Ag_2SO_4 during digestion. The amount of Fe^{2+} present in well-aerated soils are very small compared to SOC and thus, can be neglected. The soils are not ground with iron or steel equipment before analysis as metallic iron (Fe) in soil may also involve in redox reactions.

Colorimetric method

The amount of $Cr_2O_7^{-2}$ remaining after reaction with SOM may also be estimated by colorimetry after soil removal by centrifugation or filtration. The green colour due to Cr^{3+} is quantified at 600 nm wavelength and the absorbance is directly related SOC. Standard curve is also obtained by different levels of OC.

Interpretation

% Organic Carbon	Rating
< 0.40	low
0.40 - 0.75	Medium
> 0.75	High

The change in the weight of soil sample after destruction of SOM by H_2O_2 treatment also not accurate, as it does not remove SOM quantitatively.

Chapter 10

Estimation of Soil pH

The P^H of soil is an important physicochemical property of soil. It influences the suitability of a soil for a crop, availability of nutrients, microbial activity and physical property of the soil. It also indicates the need for lime and gypsum application.

The term pH was proposed by Sorensen (1909) and defined as "the negative logarithm of the H^+ (hydrogen ion) concentration".

$pH = - \log (H^+)$

Later, it was realized that the activity of an ion rather than concentration determines the emf of galvanic cell of the commonly used method to measure, pH and hence, pH may be defined as 'the negative logarithm of hydrogen ion activity'.

$P^H = - \log a H+$

The activity of an ion can be considered as the effective concentration and can be computed by,

$$a_i = c_i \times f_i$$

Where, a_i = activity of an ion 'i'
c_i = concentration of the ion 'i'
f_i = activity coefficient of the ion 'i'

$$- \log f_i = \frac{ai\ Zi_2 I}{1 + I}$$

$$I = 1/2 c_i z_i^2$$

Where, Zi = charge of an ion
a_i = activity coefficient (constant = 0.51)
I = ionic strength

Thus, mathematically pH = - log10 H^+, where, H^+ is the activity of H^+ in g ions per litre. Pure water is dissociated into H^+ and OH^- as per the following equation :

$$\frac{(H^+) \times (OH^-)}{(H_2O)} = 10^{-14}$$

Taking (H_2O) as Unity, (H^+) x (OH^-) = 10^{-14} and H^+ =10^{-7} or pH = 7. So, pure water has a pH of 7.0. As the H^+ activity increases, the pH value decreases and vice- versa. Thus, pH is a measure of the active acidity or alkalinity of water, soil or any other substance in an aqueous system. The pH scale ranges from 0 to 14. The midpoint of this scale (7.0) is neutral. The pH above 7 indicates alkalinity and that below indicates acidity.

Methods

The pH of a solution can be determined by the following methods:

I. Colorimetric method

II. Potentiometric method

I. Kuhn's colorimetric method

Principle

Indicators (organic dyes) have the property of changing colour with the changing H^+ concentration. $BaSO_4$ settles the colloidal particles in soil suspension and the quality required depends on the extent of the colloidal material. The colour of the indicator, which is not observed by the soil denotes the pH of the soil. Indicators are sensitive to H^+ concentrations and are differently coloured in solutions differing in acidity and alkalinity. The colour changes associated with structural re-arrangements. The colour change of an indicator takes place gradually, since it depends on the ratio of the concentrations of the acidic form and the basic form of the indicator.

The range of pH values, within which the indicator changes colour is different for different indicators is shown in the table below.

S. No.	Indicators	Colour Change	pH Scale
1	Thymol Blue	Red-Yellow	1.2-2.8
2	Bromophenol Blue	Yellow-Violet	2.8-4.5
3	Bromocresol Green	Yellow-Blue	3.6-5.2
4	Methyl Red	Red-Yellow	4.2-6.3
5	Bromocresol Purple	Yellow- Purple	5.2-6.8
6	Bromothymol Blue	Yellow-Blue	6.0-7.6
7	Cresol Red	Yellow-Red	7.2-8.8
8	Thymol Blue	Yellow-Blue	8.0-9.6

Reagents and apparatus required

a) Barium sulphate
b) Universal indicator
c) pH chart and
d) Test tubes

Procedure

1. Place about 1 cm thick layer of $BaSO_4$ at the bottom of the test tube
2. Place 3 cm thick soil above the first and add distilled water (DW) about 9-10 cm
3. Add 0.5 ml of universal indicator and fill the tube with DW to a height of 15 cm
4. Plug the tube with a cork and shake vigorously for about a minute and allow for settling
5. Compare the colour of suspension with the chart on the pH bottle
6. Select the disk on which the above estimated pH value is written and from this disk, find indicator required to know the exact pH
7. Repeat the steps from 1 to 5 using the indicator selected in step 6
8. To know the exact pH of the given soil, compare the colour of the suspension in the presence of the above indicator with the colour standards fitted into the Lovibond comparator disc.

II. Potentiometric method

The activity of H^+ is calculated from the measured potential of half-cell involving this ion, and the known standard potential of the half-cell.

Suppose a metallic rod is dipped in a solution of one of its salts, it is found to acquire an electric charge, which reaches maximum value after some time. This is due to the solution or the metal rod takes ions from the solution and this may lead to the development of an electrical potential at the metallic rod known as electrode potential. This potential according to Helmholtz, is developed due to two opposing forces acting in the half cell, namely the solution pressure and the osmotic pressure.

Nernst has shown that the potential difference between the metal and the solution of its salt is given by the equation:

$E = E^o - RT/nF \log_e P^o/P$

Where, E= Electric potential (observed)

E^o= Standard electrode potential at unit activity

R= Gas constant (8.313 J)

T= Absolute Temperature

n= No. of electrons taking part in the reaction

P^o= the solution pressure

P= the osmotic pressure

In the reaction,

$H_2 \text{ (gas at 1 atm)} \longrightarrow 2H^+ + 2e^-$

And by considering E^o is for the hydrogen electrode, we get,

$$E = -RT/nF \log_e H^+/ (\text{Pressure of } H_2)$$

$$= \frac{-RT \times 2.303 \times \log_{10} (H^+)}{F}$$

Thus,

$pH = E/0.591$ at 25^0C

If we can find the electrode potential developed by dipping it in the solution, we can calculate pH. Such an electrode is called half-cell and is also called as "indicator electrode". It is not possible to find the emf of this half-cell practically. Hence, it is coupled with another half-cell of constant value and is called as "Reference electrode". In making pH measurements, the H electrode in the test solution and it is placed contact with Calomel electrode (reference electrode) through KCl salt bridge.

Hence,

$$E = E_{H2} - E_{cal}$$
$$= 0.0591\ pH - E_{Cal}$$

Since, the Calomel electrode exhibits a potential of 0.285 V the pH can be calculated by using formula,

$$pH = \frac{E - 0.0285 \text{ at } 25^{\circ}C}{0.0591}$$

Since, the hydrogen electrode has some difficulties, a glass electrode is used in pH measurements.

Glass electrode

It consists of a thin-walled bulb made of chemically pure and soda glass. A suitable electrolyte, usually 0.1 N HCl is present in the bulb and a silver wire coated with AgCl, is dipping in acid present inside the bulb. The potential developed at the glass membrane due to difference in the concentration of hydrogen of the outside and inside solutions is measured by connecting it to a reference (Calomel) electrode. The potential developed is measured in terms of volts (generally in mV has the potential developed is very small).

A pH metre is an instrument which measures the voltage developed by the combination of a glass electrode and reference electrode. The scale is graduated in terms of both pH and millivolts (mV). A schematic diagram of a glass assembly is given in the Fig.10.1.

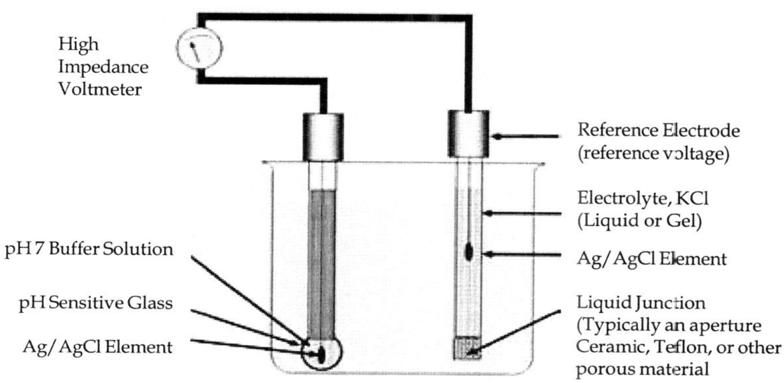

Fig.10.1: Schematic diagram of pH meter

An ordinary voltmeter cannot be used for measuring the voltage of glass electrode-reference of electrode assembly for the reason that electrode being a non-conductor allows an extremely small current, which would not activate a galvanometer. In a pHmeter, this small current is amplified through electronic tubes so that it can be measured easily.

Measurement of soil pH

pH meter with glass-calomel combined electrode is most commonly used.

Reagents

1. **pH4.0 buffer solution:** Dissolve 10.21 g of Potassium hydrogen phthalate (0.05 M) in water and dilute it to one litre. This solution should not to be used after 5 to 6 weeks or earlier, if fungal growth is seen.
2. **pH7.0 buffer solution:** Dissolve 0.948 g of KH_2O_4 and 1.57 g Na_2HPO_4 in water and dilute it to one litre.
3. **pH9.2 buffer solution:** (0.01M Borax) dissolve 3.81 g of $Na_2B_4O_7$. 10 H_2O in carbon dioxide free water dilute to one litre. The solution should be protected from atmospheric carbon dioxide and should be replaced once in a month.

The Buffer tablets or powder are available commercially for pH4.0, 7.0, 9.2. This buffer solution can be prepared by dissolving powder/ tablet in required quantity of boiled cooled distilled water (100 ml).

Procedure (for Elico pH metre)

1. Take 20g of soil in a 50 ml beaker, and add 25 ml of distilled water and stir at least 5 times a 30 minutes period to allow for soil and water to reach equilibrium.
2. In the meantime switch on the instrument, set temperature knob to room temperature and set range selector to zero.
3. Set the galvanometer to zero with set zero knob after 15 minutes.
4. Dip the electrodes into buffer to know the pH, turn the range selector and adjust the pH metre to pH of the buffer solution by set buffer knob.

5. Turn the selector to zero, press the key stand by hand and take out the electrode. Wash the electrode with distilled water and wipe it a piece of filter paper.
6. Shake the soil suspension, insert the electrodes into it. Turn the selector in the proper pH range and read the pH on the dial of the pH metre. Turn back the selector to zero and take out the electrode.
7. Rinse the electrode with distilled water and lower them back into a beaker containing distilled water.

The pH value is erratic and unreliable, when determinations are made at lower soil moisture due to it.

- Poor contact between soil and the glass electrode.
- The high mechanical resistance of the soil. On the other hand, too high a soil- water ratio increases the pH value owing to dilution. In general, more dilute the soil suspension, the higher the soil pH values found, whether the soil is alkaline or acidic.
- The rise in the pH value by 0.2 to 0.5 as soil moisture increases from the sticky point. A 1:2.5 ratio of soil to water seem to give the most uniform results.

Precautions in the use of pH metre

- The electrons should not be allowed to remain in the suspension or test solution longer than necessary.
- The electrode should be washed with gentle stream of distilled water immediately.
- The electrodes should be suspended in distilled water when it is not in use. The drying of electrodes should be avoided.
- If soil has more soluble salts, soil suspension may be prepared, using either 0.1 M $CaCl_2$ or KCl solution instead of water to overcome the effect of soluble salts on pH.

pH	Nature of soil	Interpretation
<6.5	Acidic	Require liming
6.5-8.5	Neutral	No amendments. Optimum for crops
>8.5	Alkaline	Requires gypsum for reclamation

Chapter 11

Estimation of Electrical Conductivity of Soil

Aqueous solutions conduct electricity like metals and both of them follows ohm's law. According to this law, the strength of the current (I) flowing through a conductor is directly proportional to the potential difference (E), applied across the conductor and inversely proportional to the resistance (R) of the conductor.

$I = E / R$

$IR = E$

The reciprocal of resistance is called conductance(C). If a solution is placed between two parallel electrodes having area 'A' cm^2 and distance 'l' cm resistance in ohms is given by,

$R = Pl / A$

Where P is the specific resistance with units ohm/cm. The reciprocal of specific resistance is called specific conductance or specific conductivity (K). It is the conductivity of 1 cc solution placed between plate electrodes of one square centimetre of area & kept one cm apart.

$K = L / P$ (where $p = (RA) / L$)

Therefore, $K = L/a \times L/R$ (*where* L= 1cm)

Specific conductivity = cell constant x conductance

Units of specific conductivity are mho/cm or ohm/cm at 25^0C.

Soluble salts concentration of soil is determined by measuring the cations and anions in soil water extract and measuring the electrical conductivity (EC) of soil extract.

Electrical conductivity is the reciprocal of resistance of a conductor and is expressed in mho/cm. In case of soils, the EC of water extract is proportional to its salts concentration.

Therefore, measurements of electrical conductivity are used in soil testing to estimate soil salinity. Since most soil solutions have a conductivity much less than unity, the measurements are expressed in terms of millimhos/cm ($EC \times 10^3$). The unit for expressing the conductivity in irrigation water is in micro mhos/cm ($EC \times 10^6$).

The conductivity of saturation extract is generally recommended for appraising soil salinity in relation to plant growth, the following equations describe the relationship between the osmotic pressure and the total soluble salts in the soil and soil conductivity measures.

$$\text{Osmotic pressure} = 0.36 \times EC \times 10^3$$

$$\text{Percentage of soluble salts} = \frac{0.064 \times EC \times 10^3 \times \text{saturation percentage}}{10^6}$$

However, the difficulty of obtaining saturation extracts makes the latter prohibitive for routine use, the salinity estimates based on the conductivity of dilute extractants or suspensions are more convenient for rapid determination.

Soils generally differ in their salt content, which affects their ability to grow crops. The soil containing an excess of soluble salts are saline. Excess salts interfere with water and nutrient content uptake, and in saline soils, all crops may not grow successfully. Various crops differ in their tolerance to presence of salts in the soils as follows :

High salt tolerant : Barley, sugar beet, cotton, spinach

Medium salt tolerant : Rice, wheat, sorghum, maize, cabbage, peas, tomato

Low salt tolerant : Field bean, radish, green gram.

Preparations of extractants

1. 400-1000 g of fine earth (<2 mm) into a beaker (to get 40 ml extractant).
2. Add a crystal of Thymol blue to reduce bacterial growth.
3. Add enough water to saturate.

4. Stir gently with a glass rod and add either water or soil to get saturation, which can be given as below :

   a) Tap the beaker smoothly on table top, free water should not get collected on the surface.

   b) The paste glistens, as it reflects the light.

   c) The paste flows slightly, when the beaker is held in a slanting position.

   d) The plates slides freely and cleanly off the spatula (except for clay soils).

5. Cover the beaker with a watch glass and allow to stand overnight

6. Check the above mentioned parameters (step 4) and adjust the soil condition.

7. Transfer the paste to a Buchner funnel and get the extract using a vacuum pump.

It is preferable to determine the soluble salt content of soil, when the moisture level is at field capacity. But as we cannot get sufficient quality of water at field capacity, saturation extract is preferred. Since, this method is laborious, 1:2 or 1:2.5 soil: water ratio extract is being extensively followed by most research workers.

The EC of the soil extract is measured by conductometry. The principle involved in this method is as follows :

Solutions offer resistance to the passage of electricity through them depending on the concentration of soluble salts. Higher the salt content, lower the resistance to the flow of current. The conductivity is the reciprocal of resistance and it depends on:

a) Length of solution (L) or distance between electrodes.

b) Area of the electrodes (A).

c) Nature of the salt present.

d) Temperature of the solution.

Conductance (C) = a / L = (K. a) / L

Hence, K = C x L/ a

i.e., Specific Conductance (K) = Conductance x soil constant

Apparatus

Conductivity metre or bridge

Reagent: 0.01 M KCl

Procedure: Determination of cell constant (K)

Cell constant is the ratio of the distance between the electrodes to the area of the electrode. Cell constant of given cells determined by measuring the conductivity of 0.01 M KCl solution, whose specific conductivity (1.412 ds/m at 25°C) taken as reference electrode.

Cell constant

= calculated conductance of KCl / Observed conductance of KCl

= 1.412/X

Determination of electrical conductivity

Principle

The EC of soil measured with conductivity meter known as '**SOLUTE BRIDGE**'. It is based on the principle of wheat stone bridge in an alternating current is used instead of direct current in order to eliminate the polarization of electrodes. The balance point is indicated by 'magic eye' or telephone.

A conductivity cell consists of two platinum electrodes in the form of rectangular piece fused on the one slide and covered with black spongy platinum on the other. Two wire head connect it to proper terminals on the solute bridge. A figure known as cell constant is marked on the cell.

Procedure: (Elico solute bridge)

1. Take 10 g of soil into beaker and add 25 ml of distilled water.
2. Stir the suspension intermittently and leave it overnight in order to obtain a clear supernatant solution.
3. Switch on the instrument for 15 minutes before taking reading for stabilization.
4. Dip the conductivity cell into the supernatant solution.
5. Move the pointer on the dial to get maximum area in shade in the magic eye.

6. Note the reading on the Solute bridge and calculated EC as given below.

Calculation and observations

Cell constant of the conductivity cell = X

Solute bridge readings = Y m. mhos.

EC = X xY m mhos/cm

Interpretation

EC (m. mhos /cm)	Nature of Soil
< 0.8	Normal
0.8 – 1.6	Critical for tolerant crops
1.6- 2.5	Critical for Salt tolerant crops
> 2.5	Injurious to many crops

Procedure

1. Keep the soil water ratio into suspension at 1: 2.5 (EC decreases with dilution)
2. Allow the soil suspension to stand for a sufficient time to obtain clear supernatant.
3. Adjust the temperature of the solute bridge while measuring EC
4. Dip the electrode of the cell completely into solution to get accurate reading.
5. Keep the electrodes dipped in distilled water, when not in use.

❑❑❑

Chapter 12

Estimation of Total Nitrogen in Soils

The total N content of soils ranges from < 0.02 % in subsoils to > 2.5% in peat; the surface layer of most cultivated soils contain between 0.06 and 0.5 per cent. The total N can be determined by :

1. Kjeldahl method – a wet oxidation method
2. Dumas method – a dry combustion / oxidation method.

1. Kjeldahl method

In the Kjeldahl method, organic-N (org-N) in the sample is converted to NH_4^+-N by digestion with concentrated H_2SO_4 along with digestion mixture, and the NH_4-N is determined from the amount of NH_3 liberated by distillation of the digest with alkali. The speed of the complete conversion of org-N to NH_4-N during digestion is made by adding salts to raise the temperature or by some catalysts. In Kjeldahl method, K_2SO_4 or Na_2SO_4 is used to increase the temperature of digestion and Se, Hg or Cu is used as catalyst to remove organic matter oxidation.

Nitrogen is present in soil in both organic and inorganic forms. It exists as ammonium and nitrate nitrogen in the inorganic forms. The most commonly used method for total –N determination is by Kjeldahl's method. It involves two steps :

1. Digestion of sample to convert org – N to NH_4^+-N.
2. Determination of NH_4^+-N in the digest.

The org-N is converted to NH_4^+-N by oxidizing with H_2SO_4 in presence of salts (to raise the digestion temperature) and catalysts (Hg, Cu or Se – to increase the oxidation rate). The NH_4^+-N in digest is determined by collection of NH^{+4} in boric acid or standard H_2SO_4

liberated during distillation of the digest with alkali. It is satisfactory for many soils and it does not give accurate results with compounds containing N-N and N-O linkages. Many of the soils have very little of the N is in the form of organic compounds, containing N-N OR N-O linkages. In this method, NO_3-N is lost by volatilization. Since, the nitrate N content of most of the soils is negligible, the less of NO^-_3 is ignored. In this method, salicylic acid is used to convert nitrogenous substances, nitrites and nitrates to ammonical nitrogen.

Materials: Digestion mixture (Na_2SO_4 or K_2SO_4: $CuSO_4.5H_2O$: selenium powder in the ratio of 100 : 20 : 0.05); concentrated H_2SO_4 (N free); NaOH (40 %) ; 0.1 N H_2SO_4; boric acid (4 %) with mixed indicator (0.1 g bromocresol green + 0.07 g of methyl red in 100 ml alcohol); salicyclic acid; sodium thiosulphate; kjeldahl flask; and distillation set.

If indicator powders are not available, then boric acid - indicator solution can be prepared by dissolving 20g pure boric in 700ml hot water and transfer the cooled solution to a volumetric flask containing 200ml ethanol and 20ml of universal mixed indicator solution. Mix the contents and add 0.05 N NaOH till the colour changes to reddish purple and then make up the volume to 1 litre.

Procedure

1. Transfer 2 g of soil into a 500 ml Kjeldahl flask (put soil in the form of paper packets to prevent soil sticking to the flask).
2. Add 1-2 g of salicylic acid and 20ml of conc. H_2SO_4 swirl the contents and keep it for 30 minutes at room temperature (for pre digestion).
3. Add 0.5 g of sodium thiosulphate and add 2g of digestion mixture and swirl to mix the contents.
4. Digest the contents on burner, heating slowly in beginning taking care not to spill the contents by frothing.
5. Continue digestion till the contents become colourless or light green in colour. Cool and transfer the contents to distillation flask using distilled water.
6. Place the receiving flask containing 4 % boric acid and mixed indicator taking care that the tip of the tube dips into the solution. Add 25-50 ml of 40% NaOH (till the contents become distinctly alkaline).

7. Close the lid of the flask and distill by heating.
8. Collect the distillate in the receiving flasks till evolution of NH_3 subsides. Remove the receiver and put off the burner.
9. Titrate the distillate with standard 0.1 N HCltill the blue colour disappears. One drop in excess turns the solution to pink. Run a blank simultaneously.

Calculations

1 Ltr of 1 N HCl = 14 g of N

1 ml of 0.1 N HCl = 0.0014 g of nitrogen

$$\text{Per cent Nitrogen in the sample} = \frac{(T - B) \times 0.1 \times 0.0014}{S} \times 100$$

Where, T and B are ml of standard acid for sample and for blank titration, respectively and S is the sample weight in grams.

2. Dumas method

The sample is heated with CuO at high temperature (> 600° C) in a stream of purified CO_2, and the gasses liberated are lead over hot Cu to reduce N- oxides (mainly N_2O) to N_2 and then, over CuO to convert C to CO_2. The N_2-CO_2 mixture is then collected in a nitrometre collecting concentrated alkali, which absorbs the CO_2 and the volume of N_2 gas is measured.

Chapter 13

Estimation of Available Nitrogen in Soil

The available N in soil represents a fraction of the total N absorbable by plants. Nitrogen is generally taken up in the NO_3^- from under aerobic conditions and as NH_4^+ ions under anaerobic conditions of plant growth. Nevertheless, the sum of the NH^{+4}-N and NO_3-N is generally smaller than the total N, which becomes available to the plants during their growth. The organic forms of N, which becomes available to the plants during their growth. The organic forms of N (protein, amino acids, amino sugars, *etc.*) gradually become available to plants for absorption, depending upon the nature of their combination and soil conditions.

Since, the bulk of the total nitrogen at any one time is present in organic combination, a method that takes this reserve nitrogen into account is better index of the N supplying capacity of a soil. The proposed microbial and biological methods are the laborious. Water, dilute and strong acids, and alkalis have been proposed as extractants among chemical methods for estimation of available N. The alkaline potassium permanganate method (Subbiah and Asija, 1956) is used in several laboratories in India.

Sahrawat and Burford (1982) showed that this method does not estimate NO_3^--N and suggested addition of 1.5 g of Devarda's alloy to the distillation flask in addition to alkaline permanganate. The modification includes distillation as in standard method till 50ml distillate is collected, then addition of Devarda's alloy, followed by further distillation till another 25 ml distillate is collected. The two stage distillation was suggested as Devarda's alloy was found to interfere with oxidation of organic – N by $KMnO_4$.

The alkaline potassium permanganate method (Mineralizable - N)

Principle

The alkaline $KMnO_4$ is a mild oxidizing agent and can extract the relatively easily oxidizable fractions of organic–N. A known weight of the soil is treated with an excess of alkaline $KMnO_4$. The ammonia evolved is absorbed in known volume of standard acid, the excess of which is titrated against standard alkali, using methyl red as the indicator.

Reactions

$$KMnO_4 + H_2O \xrightarrow{\text{Alkaline medium}} 2MnO_2 + 2KOH + 3\,(O)$$

$$R\,CH.NH_2.COOH + \tfrac{1}{2}\,O_2 \xrightarrow{\text{Oxidation}} R - CO.COOH + NH_3$$

$$NH_3 + H_2O \xrightarrow{\text{Distillation}} NH_4OH$$

$$NH_4^+ + 2NaOH + OH \longrightarrow NH_4OH + 2NaOH$$

$$H_2SO_4 + NH_4OH \xrightarrow{\text{Absorption}} (NH_4)_2SO_4 + 2\,H_2O$$

$$H_2SO_4 + NaOH \xrightarrow{\text{Titration}} Na_2SO_4 + 2\,H_2O$$

Reagents and apparatus required

1. 0.32% $KMnO_4$: Dissolve 3.2 g of $KMnO_4$ in distilled water and make up the volume to 1 litre.
2. 2.5% NaOH: Dissolve 25g of NaOH flakes in distilled water and make up the volume to 1 litre.
3. 0.02 N H_2SO_4
4. 0.02 N NaOH
5. 0.15 % Methyl red indicator: Dissolve 0.15g of methyl red powder in 100ml alcohol.
6. A Kjeldahl distillation unit
7. Measuring cylinder
8. Distillation flask
9. Pipette, burette and 250 ml conical flask

Procedure

1. Weigh 20g of soil on to a filter paper, prepare a packet and put it into the Kjeldahl distillation flask and fix the flask in the Kjeldahl assembly.
2. Add 100ml of 0.32% $KMnO_4$ solution add 100ml of 2.5% NaOH solution and fix the cork immediately. Pipette out 25ml of 0.02 N H_2SO_4 in conical flask and add 3-4 drops of methyl red indicator and dip the end of the delivery tube into it.
3. Distill the ammonia gas from the distillation flask with the help of burner and collect about 30ml of the distillate (till no ammonia is liberated).
4. Titrate the excess sulphuric acid in conical flask against 0.02 NaOH till the colour changes from pink to yellow and note down the volume of NaOH used.

Precautions

1. Before adding NaOH solution to the distillation flask. The delivery tube must be dipped into the dilute acid contained in the receiver.
2. Immediately after the addition of NaOH, the mouth of the flask should be closed, so as to avoid loss of ammonia produced as a result of the alkaline reaction.
3. The indicator should be added before starting distillation to ensure that the receiver contains enough of acid.
4. The connections should be air tight so that liberated NH_3 is not lost.
5. Before putting off the burners, the delivery tube end must be removed from the receiver to avoid 'back suction', when the distillation is over.
6. Titration should be done carefully not to cross the endpoint.

Observations and Calculations

Weight of the soil taken	:	20 g
Volume of 0.02 N H_2SO_4 taken	:	25 ml
Volume of 0.02 NaOH used	:	X ml
Volume of 0.02 N acid used for NH_3	:	(25-X) ml

(1 ml of 0.02 N H_2SO_4 = 0.00028 g of N)

$$\% \text{ available N} = \frac{(25\text{-X}) \times 0.00028}{20} \times 100$$

Available N in ppm = % x 10,000

Available N in kg/ha = % x 22,400

Interpretation

Amount of nitrogen (kg/ ha)	Rating
< 280	low
280-560	medium
>560	high

Recent methods to determine available N

In recent years, over methods to assess N fertility status have received attention. They are (a) Profile NO_3^- level, (b) mineral N (NH^+_4-N) released under aerobic incubation (Kenny, 1982) and anaerobic incubation method (Subbaiah and Bajaj, 1962; Warring and Bremner, 1964). In the profile method, a sampling depth of 60 cm and collection of about 20 core samples per field are recommended. The soil should be air dried or frozen soon after sampling and the NO^-_3-N estimated. The main limitation of this method is that the NO^-_3-N levels fluctuate during the season, as it was subjected to leaching, de-nitrification, and immobilization by micro-organisms. The anaerobic incubation method is simple, reliable and readily adoptable to routine analysis, requires no elaborate equipments, glassware and chemicals. Work done at IARI showed that this method serves as a multi-nutrient soil test. The only limitation is, it requires at least 7 days to obtain the test value.

Aerobic incubation method

The soil sample is kept for incubation under aerobic condition for a known period and the amount of N extracted by KCl is quantified.

Procedure

1. Weigh 10 g soil and mix with 30 g of acid washed quartz sand (30-60 mesh) and transfer to a 250 ml wide mouth bottle containing 6.0ml distilled water. Mix thoroughly and distribute the material evenly on the bottom of the container.

2. Close the bottle with a rubber stopper having a hole at the center through which insert a glass tube to provide aeration. Incubate the sample at 30 °C for 14 days.
3. Remove the stopper, add 100ml of 2 N KCl and close the mouth of the flask with a solid rubber cork and shake the contents for 1 hour on a mechanical shaker.
4. Filter the contents using Whatman No. 1 filter paper.
5. Take 20 ml clear filtrate into a distillation flask, add 80 ml distilled water and 0.2 g each of fresh MgO and Devarda's alloy and distill at moderate speed. Collect the distillate in 20 ml of 4% boric acid solution.
6. Titrate the liberated ammonia against Std. acid. Deduct the amount of NH_4^+-N, NO_2^--N and NO^-_3-N present in the soil without incubation from the values obtained with incubation to obtain the quantity of N released.

Procedure for anaerobic incubation method

The soil is incubated under anaerobic (water-logged) condition for a known period and the ammonia liberated during this period is quantified.

Procedure

1. Take 12.5 ml of distilled water in 16 x 150 mm test tube or 150 ml conical flask.
2. Add 5 g of air dry soil and close the mouth of the tube or flask with a rubber stopper and incubate at 40°C for 7 days.
3. Transfer the contents to a 250 ml conical flask, add 25 ml of 4 N or 50 ml of 2 N KCl, shake for 30 minutes mechanically and filter.
4. Take a known volume of aliquot and distil by adding 0.5 g MgO. Collect the distillate in 20 ml of 4% boric acid, and titrate the ammonia against Std. acid.
5. Determine the ammonia content of the soil without incubation and deduct from the amount obtained after incubation.

Chapter 14

Estimation of Ammonical and Nitrate Nitrogen in Soils

I. Determination of ammonical-N in soils (Nessler's method)

The NH_4^+ in soil is subjected to oxidation to NO_2 to NO_3 (nitrification) by soil microbes forms in soils, hence its determination should be done quickly after collecting the soil samples. Exchangeable NH_4^+ can be determined by several methods, but they are tedious compared to other colorimetric method. The NH_4^+ content of the soil usually ranges from 1.4 to 14.0 ppm and Nessler's procedure is usually followed for greater sensitivity.

Reactions

$$HgI_2 + 2\,KI \longrightarrow K_2HgI_4$$

$$2\,K_2HgI_4 + 2\,KOH + NH_3 \longrightarrow Hg_2O\,(NH_2I) + 7\,KI + 2\,H_2O$$

Reagents

A. **Extractant solution**: 10% NaCl or KCl acidified to pH 2.5 with HCl.

B. **Nessler's reagent**: Dissolve 45.5 g of mercuric iodide and 35 g of KI in a few ml of water. Then add 112 g of KOH and bring the volume to 800 ml. mix well and cool and dilute to 1 litre with water. Allow the solution to stand for few days and decant the clear supernatant into an amber coloured bottle.

C. **Std. NH_4Cl solution**: Dissolve 1.490 g NH_4Cl in 1 litre of water to get 500 ppm of NH_4^+. Prepare a working standard having 20 ppm concentration.

D. 10 % sodium tartrate solution.

E. 1 % gum acacia solution.

Procedure

Extraction of ammonium

1. Weigh 100 g of fresh soil sample into a 500 ml conical flask, add 200 ml of extractant and shake intermittently for 15 minutes.
2. Filter through a Whatman No. 42 filter paper and wash the soil with another 250 ml of acidified extractant. Collect the leachate and make upto 500 ml.

Development of colour for standards and for unknown samples

1. Pipette out 2.5, 5.0, 10.0, 15.0, 20.0 and 25.0 ml of 20 ppm Std. solution into a series of 100 ml volumetric flasks. Add 2 ml of 10 % sodium tartrate solution of 100 ml of gum acacia solution.
2. Add an equal volume of acidified extractant, then add water to make up the volume to about 93 ml and shake the contents.
3. Add 5 ml of Nessler's reagent, mix rapidly and make upto 100 ml.
4. Read the intensity of colour at 410 nm using spectrophotometer after 25 minutes.
5. Plot the Std. graph.
6. For soil, take a known volume of the extractant and follow the steps 2 and 4.

Calculations

$$NH_4\text{-N (lbs/ac)} = \frac{\text{Graph ppm x volume made x volume of extractant x 2}}{\text{Aliquot taken x weight of soil}}$$

II. Determination of nitrate-N in soils (Phenol disulphonic acid method)

This method of estimation of nitrates depends on the nitration of 2, 4-disulphonic acid in fuming sulphuric acid.

$$C_6H_3OH\ (HSO_3)_2 + 2HNO_3 \longrightarrow C_6H_2\ (OH)\ (NO_2)_3 + 2H_2SO_4 + H_2O$$

The product trinitrophenol behaves as a nitrophenolic type indicator and is colourless in acid medium and yellow, when neutralized or in alkaline solution. A hydroxide, such as KOH or NH_4OH is employed to shift the pH to the yellow range.

$$C_6H_2\ (OH)\ (NO_2)_3 + 2\ KOH \longrightarrow C_6H_2\ (OK)\ (NO_2)_3 + H_2O$$

The nitrate solution is dried out prior to determination as the reaction is affected in the virtual absence of water.

Reagents

A. **Extractant solution**: Mix 20 ml of 1 N $CuSO_4$ and 100 ml of 0.6 % $AgSO_4$ solution and dilute to 1 litre.

B. **Phenol 2, 4 disulphuric acid**: Dissolve 25 g of pure phenol (white in colour) in 150 ml of conc. H_2SO_4. Then add 75 ml of conc. H_2SO_4. Mix the solution and heat in boiling water for 2 hours.

C. **Std. NO_3 solution**: Dissolve exactly 0.7221 g of pure KNO_3 in water and dilute to 1 litre to give 100 ppm of NO_3. Prepare a dilute NO_3 standard having 10 ppm concentration.

D. **6 N NH_4OH solution**: Mix one part of strong ammonia in one part of water.

Extraction of NO_3 from soil

1. Weigh 50 g of soil into a 500 ml conical flask, add 250 ml of extracting solution, shake for 10 minutes and add 0.5 g $Ca(OH)_2$. Continue shaking for 5 minutes and add 1 g $MgCO_3$ (to precipitate Ag and Cu and clarify the solution).
2. Collect the filtrate by filtering through a dry Whatman No. 42 filter paper.

Estimation (development of colour)

1. Take 20 ml of aliquot in a porcelain dish and evaporate to dryness.
2. Cool and add 3 ml of phenol disulphuric acid directly into the center of the basin (use rubber bulb with pipette)
3. Swirl the content so that all the residuals salts come in contact with phenol disulphuric acid, and allow the reagent to act for 10 minutes.
4. Add 15 ml of cold water and stir the solution with a glass rod.
5. Add 6 N NH_4OH slowly till the solution turns yellow (by alkaline condition) and add 3 ml excess (care: total volume should not exceed 100 ml).

6. Transfer all the contents of the basin to 100 ml volumetric flask.
7. Read the intensity of yellow colour at 420 nm using a spectrophotometer.

Preparation of Standard graph

1. Pipette out Std. solution to get 1, 2, 3, 4 and 5 ppm of NO_3 into a porcelain dish and evaporate to dryness. Follow the steps 2 to 7 indicated above.
2. Plot the absorbance against concentration and obtain a standard graph.

Calculations

$$NO_3\text{-N (lbs/ac)} = \frac{\text{Graph ppm x volume made up x volume of extractant x 2}}{\text{Aliquot taken x weight of soil}}$$

Chapter 15

Estimation of Available Phosphorus in Soil

The available P in a soil represents a fraction of the total-P, which is susceptible to plant uptake during their growth. Plants take up P in $H_2PO_4^-$ and HPO_4^{-2} forms, but the water soluble P at any one time is very small. P exists in soil in both organic and inorganic forms. The organic fractions of P (phytins, nucleic acids and inositol phosphates) do not contribute to the P nutrition of plants in our soils, as they are low in organic matter. Among the inorganic fractions Ca, Fe and Al-phosphates are in a dynamic equilibrium with water soluble P. Therefore, a measure of these inorganic P fractions is considered a useful index of the available P in the soils.

There are two distinct phases in soil P determinations:

1. Extraction of P (Bringing the soil P into the solution).
2. Quantification of the extracted P

The choice of the extractant depends on the nature of the soil and the kinds of P compounds found in the soil. First of all, a preliminary trail has to be conducted to get information on the nature of the extractant to be used for a particular soil, as the nature of the phosphate compounds vary with the type of soils. As different extractants extract different types of P compounds present in the soil, selection of proper extractant is essential. Also, there should be proper correlation between soil test values and crop uptake.

The various methods of available P extraction and estimation are:

1. Bray's No. 1 method (Bray and Kurtz, 1945).
2. Bray's No. 2 method.
3. Olsen's method (Olsen *et al.*, 1954).
4. Modified Olsen's method (Hunter, 1972).

5. Trough's method (Trough, 1930).
6. Mehlich's method (Nelson *et al.*, 1953).

In addition to the above methods of P estimation, the other methods available are :

1. Ascorbic acid method (Watanabe and Olsen, 1965).
2. Laverty's method (Laverty, 1962).

The determination of available P is mostly done by Bray's method for acid soils and Olsen's method for neutral, alkaline and calcareous soils and is normally followed for soils having high CEC and high base saturation with very low free $CaCO_3$. The modified Olsen's method is appropriate for soils having pH more than 6.5 and in areas, where soils have a wide range of pH (both below and above 6.5 pH).

I. Bray's No. 1 method (widely for acid soils)

Principle

The combination of HCl and NH_4F is designed to extract easily acid-soluble forms of P, largely Ca-phosphates, Al and Fe phosphates. The NH_4F dissolves Al, Fe and Mn-phosphates by forming complex ion with these metal ions in acid solution (in soluble compounds of Al, Fe and Mn fluoride are formed) and thereby releases phosphate ion into the solution. In presence of chloromolybdic acid in an acidic medium, the phosphate ion forms a heteropoly complex compound of P, on reduction, imparts blue colour to the solution. The intensity of blue colour can be measured at 660 nm, using a spectrophotometer.

Reaction

$$Al_3\text{-}PO_4 + 3NH_4F + 3HF \longrightarrow (NH_4)_3Al\,F_6 + H_3PO_4$$

$$H_3PO_4 + 12\,H_2MoO_4 \longrightarrow H_3\,P(Mo_3O_{10})_4 + 12\,H_2O$$

Apparatus

Photoelectric colorimeter, shaker (end to end).

Reagents

1. **Bray's extractant No. 1 (0.03 N NH_4F in 0.025 N HCl) :** Dissolve 1.11 g NH_4 in about 250 ml distilled water and add 2 ml of conc. HCl and make up the volume to 1 litre.

2. **1.5 % ammonium molybdate :** Dissolve 15 g of ammonium molybdate in 300 ml of hot (50-60°C distilled water, filter, if there are any sediments). Cool and add 350 ml of 10 N HCl slowly with rapid stirring. Cool and make upto 1 liter and store in an amber coloured bottle (If HCl is conc. and calculated amount of HCl equivalent to 350 ml of 10 N HCl).
3. **Stannous chloride stock solution (40%) :** Dissolve 10 g of $SnCl_2$ in 25 ml of conc. HCl add a piece of pure metallic tin and store in an amber coloured glass stoppered bottle. Its working solution is prepared by diluting 0.5 ml of the stock solution to 66 ml with distilled water, just before use. Prepare fresh working solution every day.
4. **Darco-G 60 (P free activated charcoal powder) :** If it is not available, P- free activated charcoal can be prepared. Take 80 g charcoal and obtain a slurry with distilled water. Leach this slurry overnight with 1 litre of 6 M HCl in 60 mm dia columns. Then leach with de-ionized water till the leachate is free of chloride. Dry the material at 110°C (Sariullah *et al.*, 1990).

Procedure

1. Weigh 5 g of soil into a 250 ml conical flask.
2. Add 50 ml of Bray's extractant, mix and shake on end-to-end shaker for 30 minutes and filter (1:10 or 1:50 can be followed). The original shaking was for 1-5 minutes, now long shaking is preferred.
3. Prepare a blank in which all reagents are added similarly.
4. Take 5 ml of the filtrate into 25 ml volumetric flask, and add 5 ml of ammonium molybdate solution. Mix well until the evolution of CO_2 ceases. If more aliquot has to be taken for analysis (as in case of low P soils) fluoride interference has to be eliminated by adding 75 ml of boric acid (5% solution) to 5 ml extract.
5. Add about 10 ml of distilled water washing the neck of the flask to remove the adhering molybdate.
6. Add about 1 ml of working stannous chloride solution and make up the volume to the mark with distilled water.
7. In the meantime switch on the colorimeter and adjust the wavelength to 660 nm. Allow to warm up for about 10-15 minutes.

8. Adjust the pointer on the galvanometer to read '0' transmission.
9. Rinse the colorimeter tube (cuvette) with the blank solution, then pour a suitable volume. Place the tube in the cuvette and turn the 100 set knob to read 100.
10. Pour the test tube solution into another colorimeter tube and read the percentage transmission or the absorbance.

Preparation of standard curve

A standard curve shows the relationship between the concentration of an element in a given solution and the intensity of its colour.

To prepare a standard curve for P, prepare standard solution of P by dissolving 0.02195 g of KH_2PO_4 in 1 litre of water (5 ppm of P solution). Pipette out 0.5, 1, 2, 3, 4 and 5 ml of this solution in different 25 ml volumetric flasks. Then proceed to develop colour as done for the test sample and record the absorbance or transmittance. Plot the absorbance values on Y-axis and concentration on X-axis and draw a line to pass through maximum no. of points. From the graph, calculate the concentration 'C' at OD of 1.0 and is used in further quantification of soil P.

II. Olsen's method

This method is mostly used for neutral, alkaline and calcareous soils.

Procedure

1. Weigh 5 g of soil into a 250 ml conical flask.
2. Add a pinch of P free Darco-G 60 and 50 N $NaHCo_3$ solution (soil: solution: 1:10 or 1:20).
3. Shake, filter and develop colour as done in Bray's method.

Precautions

1. Test tube (cuvette) should be handled from the top rather than the lower part and it should be wiped clean with a piece of tissue paper before putting it in the cuvette holder.
2. The filtrate should be colourless, if not, add more of Darco-G, shake and filter.

3. Working stannous chloride should be prepared a fresh after 2-3 hours.
4. The blue colour starts fainting after about 15 minutes, hence, the reading should be taken within 10 minutes of developing colour.
5. If the intensity of colour is too high, take a smaller amount of aliquot and develop colour. Remember that the colred solution cannot be diluted.
6. Temperature and shaking speed increases the P extracted by Olsen's extractant. So, they have to be regulated.
7. For quality analysis, a minimum of one reference and one blank should be included in a batch of 25 samples. Duplicates are done on approximately 5% of samples.

Observations and Calculations

Weight of the sample	: 5 g
Volume of the extractant	: 50 ml
Volume of aliquot (filtrate) taken	: 5 ml
Final vol. after development of colour	: 25 ml
Dilution factor	: (50/5) x (25/5) =50 times
Transmittance of the test solution	: T
Conc. of P read from the Std. curve	: X Ppm
Ppm of available P in the soil	: X x50
Kg of available P per ha	: ppm x 2.24

OR

Available P_2O_5

$$= \frac{\text{Graph ppm x Volume of extractant x Volume made x 2x 106 x 2.2}}{10^6 \text{xWeight of soil x Aliquot}}$$

Interpretation

Amount of P_2O_5 (kg/ha)	Ratings
<22.9	Low
22.9-56.33	Medium
>56.33	High

Other methods for P-Extraction

I. Mehlich's method

Reagents

1. Mehlich's extractant: 0.05 N HCl + 0.025N H_2SO_4 with pH 1.0.
2. Vanadomolybdate.

Procedure

Weigh 3g soil and 21 ml of extractant and shake for 5 minutes and filter (use activated charcoal to get clear solution). Take known aliquot and add 10 ml Vanadomolybdate solution, make up the volume and read the yellow colour intensity at 420 nm wavelength.

II. Bray's No.2 method

Bray's No. 2 extractant is made of 0.03 N NH_4F + 0.1 N HCl. The procedure to be followed and other reagents used are same as in Bray's No.1 method.

III. Trough's method

Reagents

1. Trough's extractant: 0.02 N H_2SO_4 with pH 2.4
2. Ammonium molybdate (1.5%)
3. Stannous chloride stock solution and working solution.

Procedure

Weigh 2 g of soil to a conical flask, add 40 ml extractant and shake for 5 minutes and filter. Take a suitable aliquot (5 ml) add 10 ml ammonium molybdate, shake and then add 1 ml of stannous chloride working solution. Read the intensity of blue colour using Spectrophotometer at 660 nm.

IV. Modified Olsen's method

Reagents

1. **Extractant:** 0.5 N $NaHCO_3$ + 0.01 M EDTA + 1 g of Superfloc-127 in 10 litre. Dissolve 420 g of $NaHCO_3$ in distilled water, 37.2 g di-sodium salt of EDTA in distilled water and 1 g of Superfloc-127

in 400 ml of distilled water separately. Mix the above three solutions and make up the volume to 10 litres, adjust the pH to 8.5 with 1 N NaOH and store in polyethylene bottle.

2. **Colouring reagent A :** Take 4.0 g of Bismuth sub nitrate (or sufficient quantity of any other Bismuth compound to get 2.8 g of Bismuth) in 1 litre volumetric flask. Add 400 ml of distilled water, swirl and as rapidly and safely as possible, 120 ml of concentrated H_2SO_4 and allow to cool. In a separate container dissolve 7.5 g of ammonium molybdate in 300 ml hot water, cool and mix with cool acid bismuth solution and make upto 1 litre. Store in cool dark place as it is sensitive to heat and light.
3. **Colouring reagent B :** Dilute 150 ml of solution A to 1 liter and 1 g of ascorbic acid. This solution has to be prepared on the day it has to be used.

Procedure

1. Weigh 2.5 g soil into a 250 ml volumetric flask, add 25 ml of modified Olsen's extractant (don't use activated charcoal in any part of this procedure)
2. Shake for 5 min and filter.
3. Take 2.0 ml filtrate, add 8 ml distilled water and 10 ml of colouring reagent B. Make up the volume and read the colour intensity at 660 nm.

Additonal Methods for Developing Colour

Ascorbic acid method

Reagent A

a) Dissolve 12 g of ammonium molybdate in 250 ml distilled water.

b) Dissolve 0.2908 g of antimony-potassium-tartrate in distilled water.

Add both these reagents to 1000ml of 5 NH_2SO_4. Mix thoroughly and make upto 2 litres. Store in Pyrex glass bottle in a cool dark place.

Reagent B

Dissolve 1.056 g of ascorbic acid in 200 ml of reagent A and mix, this reagent should be prepared fresh, as and when required.

Procedure

Pipette out 5ml aliquot (Bray's or Olsen's extractant) in a 25ml volumetric flask. Acidity with 5 NH_2SO_4 to pH 5.0. Determine the volume of acid required to bring the solution to pH by using P-nitrophenol, which gives yellow colour at pH 5.0. Add 20 ml of distilled water and 4ml of reagent B. Read the intensity of blue colour after 10 minutes at 660 nm.

Laverty's method

1. **Amino-napthol-sulphonic acid reducing agent :** Mix 2.5 g of 1-amino-2 napthol-4 sulphonic acid, 5 g sodium sulphite (Na_2SO_3) and 146.25 g of sodium metabisulphite ($Na_2S_2O_5$). Grind the mixture to a fine powder and store in an air tight bottle. Dissolve 8 g of the powdered mixture in 50 ml warm distilled water and keep the solution overnight. Next day, store in a glass stoppered bottle. This solution can be used for 3 weeks.
2. **Molybdate reagent :** Dissolve 7.53 g of ammonium molybdate in 500 ml hot water (60°C). Cool and add 150 ml HCland 1 g of boric acid. Dilute to 2 litres. Store in a glass stopper bottle.
3. **Dilute HCl**: 0.42 and 0.59 N HCl and Standardize the acid.

Procedure

Pipette out 5 ml of Bray's or Olsen's extract. Add 2.5ml of molybdate reagent. With a bulb pipette add 7 ml 0.42 N HCl for Bray's extracts and 9.0 ml of 0.59 N HCl for Olsen's extracted and mix well. Add 0.5 ml of sulphonic acid reducing agent and mix again. Read blue colour intensity at 660 nm after 15 minutes.

The principle of colorimetric analysis

Colorimetry is based on the principle that coloured solutions preferentially absorb some wavelengths and the colour seen is characteristic of the wavelengths not absorbed. The colour on the wave-length band of the incident light used for a colorimetric determination is selected to coincide with that most absorbed by the test solution, and this is done by the passage of light through a light filter.

According to Beer's law, equal successive increments of concentration dC, of a true solution containing a coloured constituent absorb equal fractions, di/I of the incident light.

Mathematically,

$$T = \frac{dI}{I_0} \times 100$$

If the colorimeter gives transmittance readings, these can be converted to absorbance by the following relationship :

Absorbance = 2 - log (%T)

Chapter 16

Estimation of Available Potassium in Soil

The available K in a soil is generally the sum of water soluble and exchangeable potassium. The reserve forms of K in soils are the non-exchangeable K and the mineral K. The neutral normal ammonium acetate extracts both exchangeable and water soluble K and this interaction is considered as a suitable index of K availability in soils. As the exchangeable K of soils are removed through cropping or leaching, some of the reserve K weathers from feldspars and micas, and become exchangeable. In regions, where soils have relatively high K supplying power, the soil testing laboratories dry the soil sample for a week or more at 95°F of with low humidity to permit the release of K into an exchangeable form before the testing is done. In soils where exchangeable K significantly contributes to the K nutrition of the current crop, the extraction of soil with boiling nitric acid is recommended.

Ammonium acetate (NH_4OAC) method: (Merwinand Peach, 1951)

Principle

The ammonium ions in ammonium acetate exchange with the exchangeable K ions of the soil. In addition, NH_4^+ holds highly charged layers together just as K^+, the release of non-exchangeable K in an exchangeable form is retarded during extraction with NH_4OAC.

Reagents

1. **Standard K solution (1000 ppm) :** Dissolve 1.91 g of K8
2. Cl (analar reagent) in 1000 ml of distilled water.
3. **Neutral N Ammonium acetate solution :** Dissolve 77.09 g of amino acetate salt is not available, add 57 ml of 99.5 % glacial

acetic acid to 700 ml of distilled water and then add 69 ml of concentrated ammonium hydroxide. Dilute about 900 ml and adjust the pH to 7.0 by the addition of 3N NH_4OHor 3 N CH_3COOH and make upto 1 litre. Store in Pyrex or polypropylene bottle.

Apparatus and glassware

A balance, 250 ml conical flask, shaker, funnels, beakers, Whatman No.1 filter paper, pipette and a flame photometer.

Flame Photometer

Principle

The atoms or ions present in solution gets energy in a flame, they get excited and result in the emission of spectrum. The energy absorbed by electrons, shifts them to positions more distant from the atomic nucleus. As the electrons, shifts them to positions more distant from the atomic nucleus. As the electrons regain their state, the previously absorbed energy is remitted as electromagnetic radiations. The wavelengths of which corresponds to the quality of energy involved in the respective electron shifts, and the quantity of radiation is directly proportional to the amount of the element emitting connected in a suitable circuit. The emission spectrophotometric analysis is most useful, where a large number of soil or plant samples have to be analyzed on a routine basis.

A Flame photometer has the following components and are schematically shown in the figure :

1. **Atomizer-burner assembly:** It consists of :

   a) A suction capillary through which the sample is sucked.

      * The rays this is the measured with a galvanometer.

   b) An air pipe which helps to spray the sample in to flame and is attached to the air compressor.

   c) A fuel gas tube attached to the gas cylinder (acetylene or LPG).

   d) A drain tube through which the heavy particles of the sample are drained out.

   e) A burner in which the fuel gas burns and the flame is produced. In this flame, the element which has been sprayed is excited.

2. **Filter :** It selects the radiation of the required wavelength.
3. **Galvanometer :** It measures the light energy that has been converted to electrical energy.

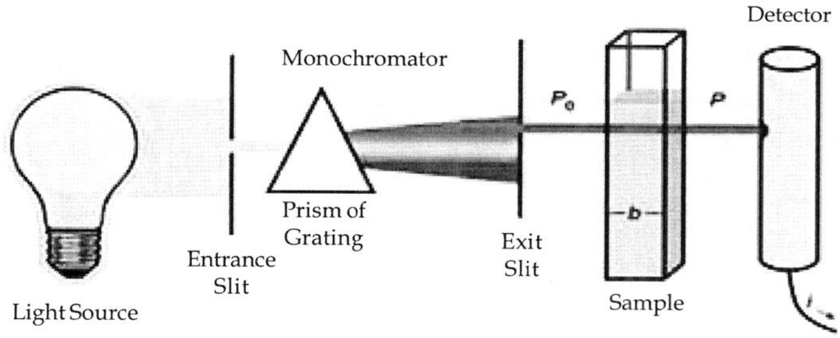

Figure 16.1: Different components of flame photometer

The following are the steps to be followed in the use of a flame photometer:

1. Connect the instrument to the mains after ensuring correct voltage.
2. Set the air compressor in operation and check the pressure gauge so that the air pressure is adjusted to 0.4 to 0.6 kg/cm^2.
3. Start gently the gas supply from the gas line fitted to the gas cylinder and light the burner carefully through the inspection window.
4. Again adjust the air and the gas pressure to give a soot free blue flame and about 4 cm in height.
5. Close the inspection window and set the galvanometer reading to zero by feeding distilled water into the atomizer.
6. Feed the standard solution of maximum concentration and adjust the galvanometer to read 100, feed other standard solution of maximum concentration and adjust the galvanometer to read 100, feed other standard solutions and note down the reading.
7. Feed the test sample and note down the reading.
8. Turn off the gas tap first, while switching off the instrument.

Procedure

1. Weigh 5 g of soil into a 250 ml conical flask.
2. Add 25ml of neutral normal ammonium acetate solution.

3. Shake the contents of the flask on an electric shaker for 5 minutes and filter.

4. Feed the filtrate in to the atomizer of the flame photometer, which has been adjusted to 100 with 40 ppm (in case of systronics flame photometer) or 10ppm (in case of Elico flame photometer) standard solution of K and note down the reading.

5. Locate this reading on the standard curve and calculate the amount of K in the soil as shown below :

Alternatively, we can place 10 g soil (5 g soil if soil has 500 ppm K) in a centrifuge tube. Add 25 ml neutral normal ammonium acetate solution and shake for 10 minutes. Centrifuge the tube at 2000 rpm for 10 minutes or till the supernatant is clear. Decant the supernatant to a 100 ml volumetric flask. In a similar manner extract 3 more times (totally use 100 ml extractant) and use this extract for determining K.

For routine soil tests extraction at 1:5 to 1:10 (soil: extractant) ratio is preferred than repeated extraction on a centrifuge. A wider ratio of 1:20 and shaking for 30 minutes, removes 95% of exchangeable K. This may be used as an alternative to centrifugation method.

Preparation of standard curve

Dissolve 1.91 g of KCl in distilled water and make up the volume to 1 liter. To give 1000 ppm solution of K from this stock solution, prepare the 100 ppm working solution from which 10, 20, 30 and 40 ppm solutions can be prepared. Adjust the galvanometer to read 100 with 40ppm or 10 ppm solution depending on the instrument, then take readings for other concentrations of K and galvanometer readings.

Precautions

1. The filtrate should be clear, to avoid clogging of the capillary.
2. The air pressure should be between 0.4-0.6 kg cm^{-2}. It should not deviate too much.
3. The gas inlet should be opened after opening the air inlet and closed before shutting off the air supply.
4. The flame should be soot free and blue.
5. The instrument should be allowed to warm up for 10-15 minutes before using it.

6. The light filter should be of the same element, which we want to determine.
7. For soils having less than 200 ppm K, calibrate the using 0-10 ppm range, standards for K contents in the range of 200-500 ppm, calibrate using 0-20 ppm range. If the K content is more than 500 ppm, the soil to solution ratio can be widened to 1:40 to 1:50.
8. When exchangeable K is relatively high, it is preferable to estimate K on field moisture samples and then calculate on air dry basis, as K is likely to be fixed during drying.

Observations and Calculations

Weight of the sample : 5g

Volume of NH_4OAC added : 25 ml

Flame photometer reading for the sample : X

ppm as read from standard curve : Y

ppm of available K in soil = Y x Total dilution
= A

$$\text{Available } K_2O \text{ (lbs/AFS)} = \frac{\text{Graph ppm}}{10^6} \times \frac{\text{Vol. of extractant}}{\text{Wt. of soil}} \times 2 \times 106 \times 1.20$$

Interpretation

Available K_2O (kg/ha)	Rating
<141	Low
141-336	Medium
>336	High

Chapter 17

Estimation of Cation Exchange Capacity (CEC) and Total Exchange Bases (TEB) of Soil

The CEC of the soil is a measure of the quantity of readily exchangeable cations neutralizing the negative charge in the soil and usually expressed in m.eq. 100 g^{-1} of soil.

Various methods proposed for determining CEC are:

1. By electrodialysis
2. Leaching soil with dilute acid
3. By summation of the replaceable hydrogen and bases
4. By replacing exchangeable cations with acetates of NH_4^+, Na^+, Ca^{2+} and $Ba^{2+}$and determining the amounts adsorbed by appropriate methods

Generally, based on saturating the soil exchange complex with differentcations, methods followed are:

1. Neutral normal ammonium acetate (suitable for acidic soils)
2. 1 N sodium acetate pH 8.2, (for alkaline and calcareous soils)
3. 0.5 N Barium chloride +0.2N triethanolamine solution (for determining CEC and also, for exchgeable H of acid soils).

Materials

Neutral normal ammonium acetate, ethanol (95%), 10% KCl solution, 4% boric acid, 0.1 N Std. HCl Mixed indicator.

Procedure

1. Transfer 10 g of air dry soil into a 250 ml conical flask, add 50 ml of neutral ammonium acetate and shake for 30 minutes and keep it overnight. Weigh another 10 g of soil into a beaker and keep it for oven drying.

2. Filter the contents using Whatman No. 1 filter paper. Transfer the residue left in the conical flask using 20 ml portions of ammonium acetate (Note: 100 ml of NH_4OAC is used for leaching). Add a pinch of ammonium chloride to the soil on filter paper or leach with 25 ml of 0.25 N ammonium chloride solution. Collect leachate and save for determining cations. Save the filtrate for determining exchangeable cations.
3. Wash the residue on the filter paper with 85% alcohol until the leachate is free of ammonia (-ve test for chlorides indicates removal of excess of ammonia). Discard the filtrate (Note: If absolute alcohol is not available, any non-polar solvents like Isopropyl alcohol can be used).
4. Wash the soil residue on the filter paper with 10% KCl solution and collect the leachate the leachate till it is free of ammonia (Cl^- test).
5. Transfer the leachate containing NH_4^+ to the distillation flask and distil ammonia by the usual procedure as done in case of nitrogen.
6. Titrate the liberated ammonia solution against Std.HCl (If KCl is not available, immediately after washing the soil with alcohol (free of excess NH_4^+ ions), it can be transferred directly into the distillation flask along with filter and distillation can be carried out).

Alternate procedure

1. Transfer 1 g of soil into a 50 ml centrifuge tube, add 33 ml of neutral normal ammonium acetate solution, shake for half an hour and centrifuge for 10 min and collect the supernatant into a suitable container. Repeat this procedure 2 more times totally using 100 ml acetate (Save the supernatant for determining exchangeable cations).
2. Add 33 ml of alcohol, shake for half an hour and centrifuge for 10 min, discard the supernatant. Repeat the step 2 or 3 times (till it is free of ammonia).
3. Then add 33 ml of 10% KCl, shake for ½ an hour, centrifuge for 10 min and collect the supernatant. Repeat this procedure 3 or 4 times or till all the ammonia adsorbed on the exchange complex are replaced by k^+ ions.
4. Distil the KCl leachate, as mentioned in the previous method.

Calculations

CEC in m.eq 100 g^{-1} of soil = T. V x N of acid x 100/S

Where,

T. V = titrate value or ml of Std. acid used

N = normality of acid

S = weight of soil

Precaution

1. The entire quantity of soil used (in the 1st method) should be transferred on to the filter paper.
2. The soil should not be allowed to drying filtration using ammonium acetate or alcohol or KCl.

Determination of CEC (m. eq 100 g^{-1}) using sodium acetate

In this method, step 1 to 4 are to be followed as in case of ammonia acetate method, then amount of sodium present in the leachate (supernatant) should be determined byusing a flame photometer (The detail of Na determination are given separately).

$$CEC = \frac{\text{ppm of Na}}{0.023 \text{ x } 10^6} \times \frac{\text{Vol. of leachate taken for dilution}}{\text{Vol. of leachate taken for dilution}} \times \frac{100}{\text{Wt. of soil}}$$

Determination of Total Exchangeable Bases (TEB)

Reagents

0.1 N HCl and 0.1 N NaOH

Procedure

Take 50 ml of ammonium acetate leachate obtained for CEC experiment into an evaporating dish and evaporate it on a water bath till it become completely dry. Ignite the residue to white ash, cool it in a desiccator and the add 50 ml of 0.1 N HCl to the dish from the sides and stir to dissolve all the salts. Back titrate the excess acid with 0.1 N NaOH using phenolphthalein indicator. Carry out blank titration.

Calculations

$$TEB = (\text{Blank - T.V of sample}) \times \text{N of acid} \times \frac{100}{\text{Aliquot taken}} \times \frac{100}{\text{Wt. of soil}}$$

Chapter 18

Estimation of Exchangable Calcium, Magnesium and Sodium in Soil

1. Determination of calcium and magnesium by titration with EDTA (Hesse, 1971)

Principle

A complexometric titration using EDTA is a classical method for determining calcium and magnesium simultaneously or individually.

Analytical success is based on :

a) Eliminating the interfering ions,

b) Achieving the required pH, and

c) Employing the appropriate indicator.

Reagents

1. **Buffer solution:** Dissolve 67.5 g of NH_4Cl in 200 ml distilled water. Add 570 ml of conc.NH_4OH and dilute to 1 litre with distilled water.

2. **Std. EDTA solution**: Dissolve 1.86 g of di-sodium EDTA (ethylene diamine tetra acetic acid) in distilled water and make upto 1 litre to get approximately 6 N solution.

3. **Std. Ca solution:** Dissolve 0.5004 g of reagent grade $CaCO_3$ (dried at 150^0 C for 6 hours) in 5 ml of approximately 6 N HCl and dilute the sodium to 1 litre to get 0.01 N solution.

4. **1 per cent cyanide solution:** Dissolve 1 g of KCN in distilled water and make upto 100 ml used to form stable cyanide of Cu and Zn. It also complexes Fe^{+2} and improves the endpoint.

5. **4 per cent potassium ferricyanide solution:** Dissolve 4 g of potassium ferricyanide (K_4FeCN_6. 3 H_2O) in distilled water and make up to 100 ml. It precipitates Mn as Mn-ferrocyanide.
6. **5 per cent Hydroxylamine hydrochloride solution:** Dissolve 5 g of hydroxyl-amine hydrochloride in distilled water and make up to 100 ml and used to reduce Fe and Mn metals.
7. **Reagent grade triethanol amine**.
8. **Buffer complex solution:** Mix 50 ml each of KCN solution, hydroxyl amine hydrochloride solution, potassium ferrocyanide solution and triethanol amine with 800 ml of buffer solution (Note: This solution not be kept for more than a week after mixing).
9. **Erichome black - T (EBT) indicator:** Dissolve 0.2 g of EBT in 50 ml of methanol (This solution should not be used after 3 weeks of preparation).
10. **Murexide indicator:** Mix 0.2 g of ammonium purpurate with 40 g of finely ground potassium sulphate and grind again in an agate/glass pestle and mortar.
11. **10 per cent NaOH solution:** Dissolve 10 g of NaOH in distilled water make upto 100 ml.

Standardization of EDTA solution

Pipette out 5.0 ml of standard Ca solution to a porcelain dish, add 5 ml of distilled water and 15 ml buffer (NH_4Cl + NH_4OH), 10 drops each 4, 5, 6 and 7 and EBT solution and stir the contents. Prepare a blank in a similar manner. Usually blank sample should be blue in colour. If not titrate with EDTA to get blue colour and note down the titer value. Now titrate the Std. Ca solution to a permanent blue colour matching the blank. Repeat the standardization using Calcium as indicator. For samples, add 10 drops of each of solution of 1-4 and 2.5 ml of NaOH solution and 1 ml of calcon solution. Titrate with EDTA till a blue colour is obtained.

Procedure

A. **Calcium + Magnesium:** 5-10 ml of ammonium extract (obtained during CEC determination) into a porcelain basin. Add 5-10 ml buffer complex to get the contents with standard EDTA till colour changes from pink to blue.

B. **Calcium:** Transfer 5-10 ml of ammonium acetate extract into a porcelain basin, add sufficient quantity of 10% NaOH solution to attain a pH of 12 or more. Add a pinch of murexide indicator and titrate it against standard EDTA solution till the colour changes from pink to violet.

Calculations

$$\text{m.eq of (Ca+mg) per 100 g of soil} = \frac{\text{T.V x N of EDTA x Vol. made}}{\text{Weight of soil x Aliquot taken}} \times 10$$

$$\text{m.eq of Mg} = \text{m. eq of } (ca^{+2} + Mg^{+2}) - \text{m. eq of } ca^{2+}$$

2. The Determination of exchangeable Na in soil

Exchangeable sodium in soils is the difference in the amount of sodium extracted by neutral normal ammonium acetate and that of water solution sodium. In the soils of humid regions, the sodium content is negligible but is appreciable in soils of arid regions, also in soils subjected to prolonged irrigation.

The sodium content of the soil can be determined using a flame photometer as in case of potassium.

Standard Na solution

Dry NaCl in an oven at 110 °C for 6 hours and weigh 2.542 g of NaCl (AR grade) dissolve in distilled water and make upto 1 litre to get a solution containing 1000 ppm of Na.

Procedure

Prepare 2, 4, 6, 8 and 10 ppm of std. solution and obtain a standard curve by adjusting flame photometer with 10 ppm solution. Feed the ammonium acetate extract (obtained from CEC determination) into the automizer of flame photometer (dilute, if necessary) and obtain the reading. Construct a standard curve and obtain the conc. of Na in the extract to calculate the total concentration either in per cent or m.eq 100 g^{-1}

$$\text{Exchangeable Na} = \frac{\text{Graph ppm x Vol. of extract x Vol. made}}{\text{106 x Wt. of soil x Aliquot Taken x m. eq wt. of Na}} \times 100$$

Chapter 19

Estimation of Available Sulphur in Soil

The most commonly used methods for estimating available sulphur in soils are :

1. Heat soluble sulphur method
2. 0.15 % $CaCl_2$ extractable S method and
3. Sodium acetate acetic acid buffer method.

1. Sodium acetate acetic acid buffer method

Principle

Available sulphur fraction of soil is extracted using sodium acetate acetic acid buffer and the liberated sulphur 'S' is already present in soil solution is precipitated by Ba^{+2}ion as $BaSO_4$. This gives turbidity to the solution which is then determined by turbidometry.

Turbidometric analysis is based on measuring the weakening of intensity of luminous flux, when it passes through a solution containing particles in suspension. The intensity decreases owing to absorption and scattering of light.

(Turbidence = $\text{Log } I_0/ I_t = K$)

The amount of available 'S' of soil is calculated, comparing the unknown sample turbidence with Standard on sulphur.

Reagents

1. **Sodium acetate + acetic acid buffer:** Dissolve 100g sodium acetate in about 500 ml water, add 30 ml of 99.5 % acetic acid make up to 1 litre.

2. **Acid seed solution (20ppm S in 6 N HCl) :** Dissolve 0.109g K_2SO_4 in one litre of 6N HCl
3. $BaCl_2$ Crystals (20-60 mesh)
4. **Stabilizing reagent :** 0.25% gum acacia solution or mix 20 ml of glycerol with 80 ml of alcohol
5. 100 ppm S solution.

Procedure

Preparation of standard curve

1. Take 0, 2.5, 5, 10, 15, and 20 ml of 100 ppm S solution in separate 50 ml volumetric flask.
2. Add 1 ml acid seed solution, 10 ml extractant and 1 ml stabilizing reagent to each.
3. Add to each 1g BaCl2 crystals and make up the volume to 50 ml.
4. Record their turbidity at 420 nm wave length using spectrophotometer.
5. Plot the turbidity against the concentrations of these solutions and draw the curve.

For soil samples

1. Add 50 ml of sodium acetate and acetic acid buffer to 10 g of soil.
2. Shake the contents for 30 minutes and filter the contents.
3. Pipette out 10 ml of the filtrate into a 50 ml volumetric flask.
4. Add 1 ml of acid seed solution and 1 ml of stabilizing reagent.
5. Add 1 g of $BaCl_2$ crystals, shake it and make up the volume to 50 ml.
6. Read the turbidity of this solution within 10 minutes at 420 nm.
7. Compare the unknown sample turbidity with Std. curve and calculate the amount of 'S' present in the given soil.

$$\text{Available S (ppm)} = \frac{\text{Graph ppm x Volume of Extractant x Volume made}}{\text{Weight of soil x Aliquot taken for turbidity}}$$

2. Heat soluble 'S' (Williams And Steinberg's, 1959)

Principle

Inorganic SO^{-2}_4-S and a part of organically held sulphates are mobilized by heat treatment and extraction is done with 1.0 % NaCl. This sulphate is converted to a turbid solution of $BaSO_4$ using $BaCl_2$. The intensity of turbidity is determined by using spectrophotometer.

Reagents

1. 1% NaCl solution: 1g of NaCl dissolved in 100 ml distilled water
2. $BaCl_2$. $2H_2O$ crystals (20-30 mesh)
3. Conditioning agent: Dissolve 75 g of NaCl in 275 ml distilled water, add 30 ml conc. HCl, 100 ml absolute ethanol and 50 ml of glycerol. Stir until NaCl dissolve and make upto 1 litre.
4. Standard SO_4 solution (given in previous method).

Procedure

1. Weigh 5 g of soil to a silica crucible and 200 ml distilled water.
2. Place the basin on a building water bath and evaporate to dryness.
3. Heat for 1 hour in a hot air oven at 102°C.
4. Cool and transfer the soil to a 50 ml centrifuge tube and extract with 33 ml of 1% NaCl repeat it twice and make use of 100 ml of 1% NaCl solution in total.
5. Pipette out 25 ml of aliquot into a silica basin and evaporate dryness with 2 ml of 3% H_2O_2 heat in an oven at 102°C for 1 hour to remove all H_2O_2. Cool and dissolve the residue in 25 ml water centrifuge to get a clear solution. Determine the S contents by taking suitable aliquot and 0.2 to 0.3 g of $BaCl_2$, shake the contents for 1 minutes and measure the turbidity at 440 nm after 2 minutes and before 10 minutes.

Useful suggestions

1. Before estimation the extract should be treated with 3% H_2O_2 (S-free) to remove interfering organic matter.
2. Wrap the rubber cork with polyethylene sheet to prevent S addition through rubber cork.

3. $CaCl_2$ extractable Sulphur (Williams and Steinberg, 1959)

Principle

S in soil is extracted with 0.15% $CaCl_2$ solution of the extracted SO_4 is estimated turbidometrically using a spectrophotometer.

Reagents

$CaCl_2$ solution (0.15%), $BaCl_2$ crystals, conditioning agent, standard S solution, as given for the previous method.

Procedure

1. Transfer 5 g of soil into a 250 ml conical flask, add 25 ml of 0.15% $CaCl_2$ solution and shake for 30 min (180 oscillations/min).
2. Filter using Whatman No. 42 filter paper
3. Take 10-20 ml extract and analyze as given for heat soluble S.

Some Suggestions

1. Instead of $CaCl_2$ solution, extraction with 500 ppm P mono - calcium phosphate solution is preferred, as $CaCl_2$ extracts only soluble S. The phosphate solution helps in extraction of adsorbed S also. The calcium phosphate is preferred, over solution and potassium, as it helps in flocculation and gives filtrate.
2. As $CaCl_2$ gives a clear extract, treatment with H_2O_2 is not necessary.
3. Field moist samples should be used as air drying increases the amount of SO_4 extracted by $CaCl_2$.

Table 19.1: Critical levels of available S (ppm) for different crops

S. No.	Crop	$CaCl_2$ extractable S	Heat soluble S
1	Rice	10	16
2	Wheat	13	13
3	Maize	13	-
4	Jowar	8	-
5	Ragi	7	-
6	Groundnut	10	11
7	Mustard	10	-
8	Sunflower	18	-
9	Soybean	14	-

Chapter 20

Estimation of Available Micro-nutrients in Soil

There are two main methods available for determination of micro-nutrient status of soil :

1. Determination by colorimetry method and
2. Determination by atomic adsorption method (AAS)

Colorimetric method is quite tedious method and for each element, separate method of extraction and estimation has to be followed. Whereas, use of AAS is quite convenient for rapid method of estimation.

Irrespective of the method followed for estimation, extraction of only the available quantity of nutrients is important. The extractant should use to extract all of the available forms of a nutrient from soils with variable properties. The amount extracted should correlate with the crops response to that nutrient under various conditions. The various extractant used for extraction of micronutrient are listed below:

1. For Iron (Fe)

A. 1N Ammonium acetate, pH 4.8: A better estimate of exchangeable Fe.

B. 0.001 M EDDHA in 0.1M $NaNO_3$: For Fe estimation in calcarious soils.

C. DTPA extractant adjusted to pH 7.3: For exchangeable Fe.

2. For Manganese (Mn)

A. 1N Ammonium acetate, pH 4.8: a better estimate of exchangeable Mn.

B. 1N Ammonium acetate + 0.2% hydroquinone: For reducible Mn.

C. 0.1N H_3PO_4: For exchangeable Mn.

D. 3N Mono ammonium phosphate: For exchangeable Mn.

E. DTPA Extractant (pH 7.3): For exchangeable Mn.

3. For Copper (Cu)

A. 1N NH_4NO_3: For exchangeable and water soluble Cu.

B. 1N Ammonium acetate, pH 4.8: For exchangeable and water soluble Cu.

C. 0.1N and 1.0 N HCl.

D. 0.02 and 0.05M EDTA: For exchangeable Cu.

E. Citrate-EDTA: For exchangeable Cu.

F. DTPA extractant (pH 7.3): For exchangeable Cu.

4. For Zinc (Zn)

A. 0.1N HCl: For occluded Zn.

B. 0.05N HCl + 0.025N H_2SO_4: For occluded Zn.

C. 0.01M EDTA + 1M $(NH_4)_2CO_3$ and 0.005M DTPA solution: For exchangeable and water soluble Zn.

D. 2N $MgCl_2$ solution: For exchangeable and water soluble Zn.

E. 1N Ammonium acetate + Dithiogone in CCl_4: good correlation with crop uptake.

F. DTPA extractant: Exchangeable Zn.

5. For Boron (B)

A. Hot Water: Ideal for B extraction.

B. Acid extractant: Not good correlation with B uptake.

6. Molybdenum (Mo)

1 N Ammonium oxalate (pH 3.3) and the various methods listed above for cationsFe, Mn, Zn, and Cu. DTPA extractant is commonly followed, as its performance was found to be better.

Determination of iron, manganese, zinc and copper using DTPA extractant

Principle

Diethylene Triamine Penta Acetic Acid (DTPA) is chelating agent which extracts the available forms of iron, manganese, zinc, and copper, when buffered. The simultaneous extraction of these elements is done by shaking the soil with DTPA for two hours. The concentration of these elements in solution can be determined by using Atomic Absorption Spectrophotometer (AAS). The AAS works on the following principle :

When a solution having a mixture of metallic species is introduced into a flame, the solvent evaporates and vapours of metallic ions are obtained. Some of the metal atoms can be raised to an energy level sufficiently high (excited state) to emit characteristic radiation - a phenomenon used in flame photometry. These atoms return to the ground state at which they absorb light energy characteristic of their own resonance wavelength (the radiation which they would emit, when they are excited). The light radiation emitting from a hallow cathode lamp (light source) absorbed by the ground state atoms will be proportional to the number of atoms in the flame. So, in AAS, the amount of light absorbed is determined as the amount absorbed is directly proportional to the concentration of the element.

Reagents

1. **DTPA extractant (0.005 M DTPA, 0.01 M $CaCl_2$ and 0.1 M Tri ethanolamine (TEA) with pH 7.3):** Dissolve 149.2 g of reagent grade TEA, 19.67 g DTPA and 14.79 g $CaCl_2$ in approximately 200 ml glass distilled water (GDW). Dilute to about 9 litres and adjust the pH to 7.3 using dilute HCl or NaOH and make upto 10 litres. This solution is stable for several months.
2. Standard solutions:
   a) **Iron standard (100 ppm Fe):** Dissolve 0.7022 g of fresh FAS (Ferrous Ammonium Sulphate) in glass distilled water and dilute to 1 litre. Prepare working standards ranging from 0 to 10 ppm. Make up the volume with the extractant.
   b) **Manganese Standard (100 ppm Mn):** Dissolve 0.406 g of $MnSO_4 . 4\ H_2O$ glass distilled water containing 1 ml of conc.

H_2SO_4 and dilute to 1 litre. Prepare the standards ranging from 0 to 10 ppm (Make up the volume with the extractant).

c) **Zinc Standard (1000 ppm Zn):** Dissolve 0.4398 g of $ZnSO_4.7H_2O$ in glass distilled water and dilute to 1 litre. Dilute this to get 100 ppm and prepare working standards ranging from 0 to 5 ppm (Make up the volume with the extractant).

d) **Copper standard (100 ppm):** Dissolve 0.393 g of $CuSO_4.\ 5H_2O$ in glass distilled water and make upto 1 litre. Prepare working standards ranging from 0 to 5 ppm (Make up the volume with the extractant).

Procedure

Preparation of the standard curve

Adjust the instrument to read zero after feeding blank solution. Then feed different concentrations of a particular element and obtain the readings and plot the values on a graph sheet and construct a standard graph.

Procedure for soil samples

1. Weigh 10 g of soil into a 150 ml conical flask, add 20 ml of DTPA extractant and shake the contents on a horizontal, shake for 2 hours.
2. Filter the suspension through Whatman No. 40 filter paper.
3. Feed the clear sample to the instrument having an appropriate hallow cathode lamp and record the readings.

Calculation

$$\text{Concentration in ppm} = \frac{\text{Graph value x volume of the extractant}}{\text{Weight of the soil}}$$

Determination of available boron content of soil

The boron is present in soil in anionic form and the amount extracted by hot water is considered as available to plants. This method was first developed by Borger and trough (1939) and then modified by wear (1965). Boron is estimated by :

1. The carmine method
2. The curcumine method

Materials required

Boron free glass wares and spectrophotometer.

1. Curcumine method

Reagents

$BaCl_2$ (10% solution), 2 M HCl solution, 0.4% $Ca(OH)_2$ solution and curcumine reagent (Dissolve 0.4g of curcumine and 0.5g of oxalic acid in 100 ml ethanol store in brown bottle in refrigerator),standard boron solution (100 ppm B:by dissolving 0.5716 g of boric acid in GDW and make upto 1 L. and phenolphthalein indicator.

Procedure

1. Reflux 20g air dry soil for 5 minutes with 40 ml of distilled water containing 0.5 ml of $BaCl_2$ solution.
2. Cool and decant the suspension to a centrifuge tube and centrifuge it to get clear solution.
3. Pipette out a suitable aliquot (to contain upto 5 ppm B) into an evaporating dish (porcelain dish), add 5 ml of $Ca(OH)_2$ solution and evaporate to dryness.
4. Cool and add 1 drop of phenolphthalein indicator and enough of 2 MHCl to discharge the colour.
5. Add 0.5 ml of acid in excess and 4 ml of curcumine reagent.
6. Heat the dish on a water bath at 55°C for 15 minutes to dry.
7. Moisten with some ethanol to a centrifuge tube and centrifuge.
8. Decant the supernatant and make upto 25 ml with ethanol and measure the absorbance at 540 nm (colour is stable for 2 hours).
9. Prepare a standard curve by taking different concentration of B and follow the steps 3-8 mentioned above.

2. Carmine method

Principle: Carmine in presence of concentrated sulphuric acid forms carminic acid, which reacts with B to form a bright red to blue coloured solution. The intensity of colour is proportional to the amount of B present in solution.

Reagents

1. 100 ppm B standard solution.
2. Conc. HCl
3. **Carmine solution:** Dissolve 920 mg of carmine in 1 litre of concentrated sulphuric acid, shake till it is completely dissolve.

Preparation of standard graph

1. Prepare a series of standard solutions containing 0.5, 1, 2, 4 and 6 ppm of 1000 ml of the above solutions into a test tube.
2. Add 2 drops of conc. HCl and 10 ml of conc. H_2SO_4. Mix and allow to cool.
3. Carefully add 10 ml of carmine reagent and mix (cover the mouth of the tube and keep for 45 minutes).
4. Measure the intensity of colour at 585 nm using a spectro-photometer.
5. Prepare a blank using 2 ml of distilled water.

Procedure of soil

1. Transfer 20g of soil into a 250 ml conical flask then add 40 ml of distilled water and keep funnel on the conical flask and boil for 5 minutes, cool and filter.
2. Pipette out 2 ml of aliquot and follow the steps 2 to 5, as mentioned for std. graph.

Calculations

$$\text{Conc. of boron in ppm} = \frac{\text{Graph ppm x vol. of water added}}{\text{Aliquot taken x Wt. of soil}}$$

Critical limits for micro-nutrient

1. Zinc 0.6 ppm (DTPA extractable at pH 7.3)
2. Copper 2.0 ppm (DTPA extractable at pH 7.3)
3. Iron 2.0 ppm (DTPA extractable at pH 7.3)
4. Manganese 2.0 ppm (DTPA extractable at pH 7.3)
5. Boron 0.5 ppm (DTPA extractable at pH 7.3)

Below the critical concentration indicated for the elements, deficiency symptoms can be seen.

Chapter 21

Estimation of Gypusm Requirement in Alkali Soils

Soil with more than 15 per cent exchangeable sodium percentage (ESP) on their exchange sites are called as alkali soils. They are characterized with poor structure and with pH more than 9.0 due to excess Na, the finer soil particles are highly dispersed, thus lowering the permeability. Such soils become unsuitable for irrigation and rendered unproductive for crops. They can be reclaimed by removing sodium present in them by exchanging with Ca^{2+} ions using calcium compounds like gypsum. For calcareous alkali soils, acid producing amendments like elemental S or pyrite are used.

In order to know the quantity of such amendments to be added for reclamation ofalkali soils, the"**GYPSUM REQUIREMENT**" of soils can be determined in the laboratory.

Principle

A known weight of soil is shaken with a saturated solution of gypsum and the decrease in the concentration of calcium in the solution is determined. This gives an index of exchangeable sodium to be replaced from soil. From this, the quantity of amendment to be added can be calculated.

Reagents

1. **Saturated gypsum solution:** Add 10-15 g of $CaSO_4.2H_2O$ to 1 L of distilled water, stir well and add some more of gypsum and stir till some amount of gypsum remains insoluble and filter the solution.
2. **NH_4Cl-NH_4OH buffer:** Dissolve 67.5 g of NH_4Cl in distilled water and add 570 ml of NH_4OH and make upto 1L.

3. **Std. EDTA solution (Versenate solution-0.01) :** Dissolve 1.861g of di-sodium EDTA salt in water and make up to 1 L and standardize it against Std.Ca solution.
4. **EBT indicator:** Dissolve 0.5 g of EBT in 100 ml of 95% methanol.

Procedure

1. Weigh 5 g of air dry soil into a 250 ml conical flask. Add 100 ml of saturated gypsum solution, shake the contents for 30 minutes and filter.
2. Pipette out 25 ml of the filtrate into a porcelain basin, add 5 ml of buffer solution and add 2-3 drops of EBT indicator. Titrate the contents against 0.01 N EDTA solution till the colour changes from wine red to bluish green. This titral value gives the concentration of Ca+ Mg in treated solution.
3. Run a blank titration with an aliquot (2 ml) of the original gypsum solution under identical conditions to be determine the concentration of calcium.

Calculations

Gypsum requirement (GR) (m.eq/100g of soil) =

$$\frac{(\text{Blank TV - sample TV}) \times \text{N of EDTA} \times 100 \times 100}{\text{Wt. of soil} \times \text{Aliquot taken}}$$

1 m eq of GR / 100 g soil
= 1.72 tonnes of gypsum per foot of soil (AFS)
= 0.32 tonnes of sulphur per acre foot of soil

Quantity of gypsum required per AFS
= GR in m. eq/100g of soil x 1.72 tonnes

Quantity of sulphur required per AFS
= GR in m. eq/100g of soil x 0.32 tonnes

Chapter 22

Estimation of Soil Acidity and Lime Requirement of Acidic Soils

A. Determination of lime content ($CaCO_3$) of soil

Calcium carbonate occurs in soils as calcite. Carbonate, which is present as a natural constituent of many soils is generally reported as calcium carbonate, even though magnesium may partly substitute calcium, resulting in the formation of Dolomite (Ca Mg $(CO_3)_2$).

The alkaline earth carbonates yield soluble Ca^{2+}andMg^{2+} which can replace exchangeable Na^+ from soil colloids. Also, it has been reported that with the increase in the concentration of $CaCO_3$, volatilization, losses of NH_3 from soils also increases. An excess $CaCO_3$ reduces the availability of some trace elements in soils by introducing too much of Ca on the soil complex.

A knowledge of $CaCO_3$ content of soil is not only imported for getting an idea about its behaviour in soil, has been mentioned above, but it is also helpful for soil survey purpose and for undertaking the mechanical analysis.

Many methods have been proposed for the determination of $CaCO_3$ in soils depending on the purpose and accuracy required. Some of the methods are given below :

1. Huchinson and Maclonnan method

In this method, soil is decomposed with cold dilute HCl in vacuum and the CO_2 evolved is absorbed in a standard solution of NaOH. The absorbed CO_2 is precipitated as $BaCO_3$ by the addition of excess of $BaCl_2$. The excess of alkali (that has not been used for the absorption of CO_2) is titrated with a standard solution of HCl in the presence of phenolphthalein indicator. From the volume of NaOH, used for absorption of CO_2, amount of $CaCo_3$ is calculated. This is the most accurate, but tedious and time consuming.

2. Rapid titration method

This method is not that accurate but very rapid compared to the previous time consuming method.

Principle

A known weight of soil is treated with a known excess of standard HCl and the unused acid is determined by back titration with a standard alkali using bromothymolblue or phenolphthalein indicator. The lime content is calculated from the volume of acid required for neutralizing lime.

Reactions

$$CaCO_3 + 2HCl \longrightarrow CaCl_2 + H_2O + CO_2$$
$$HCl + NaOH \longrightarrow NaCl + H_2O$$

Reagents

0.5 N HCl, 0.5 N NaOH and Bromothymol blue or Phenolphthalein indicator.

Procedure

1. Weigh 10 g of soil into a 250ml conical flask. Add 100ml of 0.5N HCl and close the mouth of the flask immediately with a rubber cork. Shake the contents till the reaction is complete and allow to stand for 1 hour and filter.
2. Pipette out 25 ml of filtrate and add 5 drops of bromothymol blue indicator and titrate against 0.5 N NaOH till the colour changes from yellow to blue.

Calculations

$$\text{Lime content of soil} = \frac{(25\text{-}X) \times 0.025 \times \text{vol. HCl} \times 100}{\text{Aliquot taken} \times \text{wt. of soil}}$$
$$= \frac{(25\text{-}X) \times 0.025 \times 100}{25 \times 10}$$

Where X is volume of 0.5N NaOH required in the titration for 25ml of aliquot

3. Modified passion's method

The CO_2 evolved by decomposing soil with HCl is maintained in a closed chamber and the increase in pressure is measured.

4. Puri's method: (widely used in our laboratories for soils having free $CaCO_3$)

Principle

The quantity of $CaCO_3$ is determined by titrating the suspension with 0.5 N H_2SO_4 in the presence of mixture of indicators- bromothymol blue and bromocresol green.

Calcium sulphate is used as a flocculating agent. Aluminium chloride is also added to the suspension, which on addition, acts as flocculating agent and lessens the pH of the suspension to a value at which the indicator shows their acid colour. The endpoint of this titrations the change of green colour to permanent golden yellow.

Reactions

$$AlCO_3 + 3H_2O \longrightarrow Al(OH)_3 + HCl$$

$$CaCO_3 + H_2SO_4 \longrightarrow CaSO_4 + H_2O + CO_2$$

Reagents

1. 0.5 N H_2SO_4
2. **0.1 N $AlCl_3.H_2O$:** Dissolve 8.05 g of aluminium chloride in water and make upto 1 litre with distilled water.
3. **Bromothymol blue indicator:** Grind 1 g of bromothymol blue. Dissolve it in alcohol and make upto 100 ml with alcohol.
4. $CaSO_4$ powder
5. **Bromocresol green indicator:** To be prepared like that of bromothymol blue.

Procedure

1. Weigh 10 g soil into a 250 ml conical flask. Add 100 ml distilled water, 0.5 g of $CaSO_4$ powder, 10 ml of 0.1 N Aluminium chloride and 10 drops of each of bromothymol blue and bromocresol green indicators.

2. Place the flask on a hot plate/burner and bring the contents to boil and take out. If the suspension turns green, $CaSO_4$ is present. If it turns golden yellow, $CaSO_4$ is absent.
3. Titrate the green suspension against 0.5 N H_2SO_4 till it turns to yellow. Bring it back to a boil, if it changes green colour keep on boiling and continue titration till a permanent golden yellow colour is obtained. Note down the titral value.

Calculations

1. Weight of soil = 10 g
2. Volume of 0.5N H_2SO_4 = X ml

   1ml of 1N H_2SO_4 = 0.025 g of $CaCO_3$

   Per cent $CaCO_3$ in soil = $\frac{X \times 0.025 \times 100}{\text{Weight of soil (10)}}$

   Note: If the soil has 5% $CaSO_4$ - suitable for growing citrus.

   If the soil has 10% $CaSO_4$-suitable for other fruit plants.

B. Soil acidity and lime requirement of acid soils

Acidity in soils is associated with climate and vegetation. Acid soils are mainly formed because of leaching of bases due to high rainfall and quick weathering of acidic rocks. Soil acidity is mainly due to activity of aluminium and hydrogen ions. Aluminium ions on hydrolysis gives rise to 3 hydrogen ions. Due to heterogeneous nature of soil, it acts as a buffer. There are two types of acidity in soil viz. active acidity (pH of the soil solution) and potential acidity (exchangeable).

Liming neutralizes the soil acidity and increases the base saturation of soil. When a liming material is added to soil, the H^+ in solution is neutralized, but due to hydrolysis of Al, H^+ will come from exchange complex. Hence, to know the lime requirement of soil, the potential acidity of soil should be taken into consideration. For effective liming programme, one should take into consideration (a) lime requirement of the crop to be grown (b) buffering capacity of soil, (c) time and frequency of lime application and (d) type of lime material.

Methods to determine lime requirement (LR) of soil

These methods are grouped into (a) electrical method (b) titrimetric method (c)estimation as a exchangeable Ca^{+2} and (d) measurement of volume of CO_2 commonly the buffer methods are used.

Buffer method to determine the LR of soil

Advantages

1. Combination of chemical substance that would not react unfavourable.
2. Rapid rate of reaction.
3. Results are not affected by soil properties.
4. Depression in pH would be linear with exchangeable H^+.

Reagents

1. **Buffer solution :** Dissolve 1-2 g of paranitrophenol, 2 ml triethanolamine, 3 g potassium chromate, 2g of calcium acetate, 10 g of calcium chloride and approximately 800 ml distilled water. Adjust the pH to 7.5 using dilute HCl or NaOH solution and dilute to 1 litre.
2. pH buffer solution (pH 4.0 and 7.0).

Procedure

1. Weigh 10 g soil into a50 ml beaker and add 20 ml buffer solution and shake for 10 minutes.
2. Determine the pH value, after adjusting the pH meter using appropriate buffers.
3. The LR is proportional to the depression in pH of the buffer. From the table, determine the LR of soil.

Table 22.1: Lime requirement scale for buffer method

Soil buffer pH	LR of soil (tons of $CaCO_3$*/acre)
6.7	1.6
6.6	2.2
6.5	2.8
6.4	3.4
6.3	4.0
6.2	4.5
6.1	5.2
6.0	5.8
5.9	6.4
5.8	7.0
5.7	7.6
5.6	8.2
5.5	8.6
5.4	9.5
5.3	10.1

*tons of pure $CaCO_3$ per $2x10^6$ lbs. of soil

Exchangeable Ca method

1. Weigh duplicate samples (10g) of soil into 250 ml conical flasks. Treat one of the flask with 1g $CaCO_3$ powder. Add 60 ml distilled water to both the flasks and keep them overnight with occasional shaking for the interaction of $CaCO_3$ with acid soil.
2. Then add 25ml of 0.2 N KCl solution and filter through Whatman No. 1 filter paper. Collect the filtrate into a 250 ml volumetric flasks. Wash with additional amounts of KCl (25ml at a time) to get 250 ml leachate.
3. Determine exchangeable Ca in both the leachates by versenate titration using EBT as indicator. The difference in the exchangeable H of the soil. From the exchangeble Ca value, the LR as tonnes of $CaCO_3$/acre can be calculated.

Calculations

$$\text{m. eq of Ca} = \frac{\text{T.V. x N of EDTA x vol. of leachate made up x 100}}{\text{Weight of soil x Aliquot taken for titration}}$$

Chapter 23

Critical Levels of Nutrients in Soils

Nutrients	Low	Medium	High
Organic carbon (%)	<0.5	0.5-0.75	>0.75
Available nitrogen (kg/ha)	<280	280-560	>560
Available phosphorus(P_2O_5 kg/ha)	<22.5	22.5-56	>56
Available potassium(K_2O kg/ha)	<130	130-335	>335
Ca	1.5meq/100g		
Mg	1 meq/100g		
S	<10ppm		
Zn	<0.6 ppm (0.5-1.2ppm)		
Fe	2.5-4.5 ppm		
Mn	<2 ppm		
Cu	<0.2 ppm		
B	<0.5 ppm (WS)		
Mo	<0.2 ppm		
Cl	<2 ppm(WS)		

Chapter 24

Water Sampling and Processing

Water sampling and analysis is a vital part in agricultural and environmental applications for studying the quality of water treatment process, distribution system, or source of water supply. Therefore, water sampling programme starts with collections of samples which accurately represent the characteristics of the bulk material and handled conveniently in the laboratory while still providing test results. The major source of error in the whole process of obtaining water quality information often occurs during sampling. *Over 50% of the faulty data that occur in laboratory test results aredue to sampling error rather than during laboratory analysis.* Much of what is presented here is based on standard methods for the examination of water and waste water (APHA, 1998).

Preparation for sampling

Suitability of the containers

The correct sample container type and volume should be used for the required analysis in consultation with the laboratory requirements, such as:

- Containers should be examined for cleanliness, ensuring it is strong and durable, so that it will not break in transit and that the cap does not leak, once it is secured.
- Containers must not contain any of the compounds that samples are to be analyzed for.
- Containers must be suitable for sampling the water.
- Container must be of the appropriate size.
- Containers must be high density polyethylene or glass containers with Teflon lid liners for most analysis.

- Rubber and cork stoppers must not be used so as to avoid the risk of contaminating the sample.

Decontamination of the container

Decontamination procedures involve washing and rinsing. The decontamination of the containers is of at most importance, for most parameters:

- Rinse the sampling device and the bottle thoroughly with distilled water or with the sample (discarding the rinse into the effluent).
- When some solid matter is adhered, use a brush with cleaner or something suitable to give it a pre-wash before taking the above action (*soaps are not suitable, as they are liable to remain inside the container*).
- The suitability of the cleaning procedure should be checked at regular intervals by filling a cleaned container with distilled or ion exchange water and then analyzing the water for the target analysis. If none are detected, then the cleaning procedure is shown to be suitable.
- For certain target analysis, there are preferred cleaning procedures. **For example, for nitrate analysis, the container should not be rinsed with nitric acid and for chloride analysis should not be rinsed with hydrochloric acid.**
- For most purposes, containers can be re-used, if they are cleaned with a non-phosphate detergent, rinsed with hot tap water, rinsed with dilute acid, and then re-rinsed with tap water, followed by distilled or ion exchange water and air drying.

Sampling techniques and equipment

Sampling techniques depends on the available equipment, the sampling site and the analysis to be performed. Some possible variables to be considered and decisions to be made include :

- Manual or automatic sampling equipment
- Location of the sampling site, whether indoors, outdoors, or difficulties involved
- Temporary or permanent sampler installation

- Power availability, compressed air
- Field equipment, such as:
  - Multiple samplers or automatic composite samplers
  - Chilled cool-boxes
  - Meters
  - Filtration apparatus
  - Adequate trouble-shooting equipment for small repairs

Recommended sampling procedure

The recommended procedure for collecting water samples is as follows:

- Water samples must be put in labeled suitable container or clear glass (label should be written in waterproof ink) depending on laboratory requirement. Ensure that the bottle is adequately labelled, so it can be identified at all stages of the transport and analytical process.
- The laboratory requirements for analysis are a minimum of 500 ml depending on required analysis (one litre would be suitable).
- Rinse the bottle at least 4 times with the water to be sampled.
- Do not dispose off the rinse water, where it may contaminate or mix with that which is to be sampled.
- Take 5 sub-samples (100-200 ml each) to make up the full sample.
- Fill the bottle to the top, leaving little or no air space and seal tightly with the cap.
- All information about water samples is recorded on the sample container and in the field log book, e.g., the number with the site code, time and date of collection, parameter to be tested, sampler's name, and other details to avoid mix-ups. Each sample is given a laboratory number.
- Sketch your field. Diagram it the way you sampled it. Be sure sampled areas are labelled the same as sample containers. (This is so you have a record of which recommendations apply to which areas – do not rely on your memory).

- Information sheet should be clearly written with copying pencil. Fill out the information sheets. The more information you can provide with each sample, the better your recommendation will be.
- Use plastic bottles for inorganic and metals, and glass bottles for organics.
- Sterile containers for bacteriological analysis.
- Filtering for metals, if a preservative is being used.
- Every time you collect a sample, you should record it on the field data sheet.
- Clearly mark and identify the separate sample containers from each location.
- Certain observations and measurements before sending it to the laboratory. These observations should include appearance (colour, haziness, turbidity, oil on the surface and odour). Also, some measurements are valid only on a fresh sample (temperature, dissolved oxygen and pH).
- Containers must be closed until ready to analysis.
- All water samples should be stored in chilled cool-boxes prior to dispatch to laboratory.

Water sampling processing

Representative sample

A representative sample is taken from a large body of water to make up a sample (sub-samples of 500 ml), which represents the whole body of water. To minimize the effect of natural variation in the data collected, the sample should be collected:

- From exactly the same location
- At approximately the same time each day
- In the same way
- Analyse them in the same way

Water samples can be taken directly from the body of water (stream or dam), or if a pump is used, from the first outlet along the supply line. Below, some instructions for taking a representative of water samples:

- **Stream**: Sample taken from main stream flow.
- **Stock drinking water**: Sample taken from near the edge.
- **Dam**: If water is pumped from a dam, and the sample is not taken from the outlet side of the pump, take the sample from near where suction side of the pump draws water. If wishing to assess changes in quality throughout the dam, where the water may settle into well-defined layers, it may be necessary to sample the chosen layers separately. The practicality of sampling from specific depths needs to be considered. The sample is taken from upper layers first.
- **Underground source**: When water is drawn from an underground source, i.e. well or a borewell, this is often the only way to sample. Allow the pump to run for sufficient time to flush out water which has been in the pipe, then take samples at time intervals of 5-10 minutes, from the first off take point, e.g., tap, trough, or sprinkler head.
- **Reticulated source:** If the water is drawn from a reticulated source, e.g., irrigation scheme, city or town water supply, check with the local supplying authority, as they may have analytical data on the chemical and other characteristics of the water.
- **Water irrigation**: Where the water is used for irrigation or piped for livestock or domestic purposes, the simplest approach is normally to let the pump do the sampling.

Types of water sampling

The two types of samples most commonly obtained are grab samples and composite samples:

- *Grab samples* are single samples collected at a specific spot at a site over a short period of time not exceeding 15 minutes. Thus, they represent only the composition of its source at the time and place of collection. A grab sample is required for test parameters which must be measured immediately after collection and for tests which require the entire contents of a single sample container for analysis or for tests, where the parameter will change over time. Examples of test parameters which require grab samples are temperature, pH, residual chlorine and dissolved oxygen.
- *Composite samples* are obtained by taking an appropriate number of grab samples collected at equal intervals or

proportional to flow. Flow proportional composite samples are collected, when the flow and waste characteristics are continually changing. This single sample reflects the average conditions of a point-source during an interval. The nature of the composite sample requires the tested parameters to be stable in the container for the duration of sampling, often 24 hours (this is usually means that the samples are refrigerated to 4°C).

Protecting and preserving samples

The physical and chemical characteristics of a water sample can change during transport and storage. Therefore, protecting and preservation methods only slow chemical and biological changes that inevitably occur after collecting sample. Some parameters change more quickly than others, and so need to be analysed on site.

Protecting and preservation of samples is the main thing to obtain good quality results:

1. The most common preservative approach is **cooling to 4°C**. This can be achieved by making slurry of ice and water and placing the collected sample in the slurry. This serves to slow biological activity in the sample and keep dissolved gases in solution. However, it can hasten certain physical processes, such as precipitation of metals.
2. Other common preservatives (prevent known chemical reactions and maintain a target analyte in solution) include:
   - Use of sodium thiosulphate (final concentration is 0.008%) to remove chlorine.
   - Use of copper sulphate or mercuric chloride to stop biological activity.
   - Use of zinc acetate to trap sulphides.
   - Appropriate pH adjustment to either, over 12 SU (Standard units) with sodium hydroxide (NaOH) or less than 2 SU with hydrochloric acid (HCl), sulphuric acid (H_2SO_4) or nitric acid (HNO_3) (depending on the intended analyse) can inactivate biological processes.
3. The analyst must be aware that acidification of nitrate-nitrite samples with HNO_3 or preservation of SO_4 samples with H_2SO_4

will invalidate the analysis, so the sampler must make adequate notation.

4. Samples which are not preserved must be analysed immediately.

Handling and storage of samples for analysis

- As soon as the water samples are received at the water analysis laboratory, they should be checked with the accompanying information list (including sample number and date of sampling should be written on the container. Information regarding samples should be entered in a register and each sample be given a laboratory number.
- The sample(s) should be kept refrigerated or on ice in a cooler (but not allowed to freeze) as soon as possible. Changes caused by growth of micro-organisms are best slowed by cold and dark.
- In general, the shorter the time between collection and analysis, the more reliable the results will be.

Health and safety

- Laboratory personnel, who are routinely exposed to water samples, that may contain live micro-organisms are encouraged to protect themselves from water-borne illnesses by wearing clean disposable gloves and washing their hands frequently.
- Care must be taken when handling hazardous reagents by using protective clothing and safety goggles.

Chapter 25

Estimation of pH in Irrigation Water

Water is essential for all aspects of life, domestic use, industrial processes and agricultural production. Its suitability for a particular purpose can be determined by analysis. Obtaining good results depends to a great extent on the following factors:

- Ensuring that the sample taken is truly representative of the water under consideration
- Using proper sampling techniques
- Protecting and preserving the samples until they are analysed (in a timely manner)
- The water and waste water should be checked for chemical, physical (turbidity and solid matter) and biological contamination before used it, to avoid environmental pollution (most of diseases can be spread to plant, animal, and human by water contaminated).

pH is a measure of how acidic or alkaline a solution is; or the power or potential of the hydrogen$[H^+]$ ion.pH stands for the negative logarithm of H^+ or in short H^+ - ion activity in mol/L.

$$pH = -\log [H^+]$$

The underlying reason is the slight dissociation of water molecules:

$$2\,H_2O \rightleftharpoons H_3O^+ + OH^-$$

In pure water at room temperature, the product of $[H^+] \times [OH^-]$ is equal to 1×10^{-14} mol^2/L^2.

Thus, the pH of pure water is:

$$pH = -\log (1 \times 10^{-7}) = -(-7) = 7$$

Keep in mind that the logarithmic scale includes a factor of 10 in regard to changes in concentration of H ion. For example, a solution which has a pH of 3 contains 10 times as many H^+ ions as the same volume of a solution with a pH of 4, 100 times as many as one with a pH of 5, a thousand times as many as one of pH of 6, and so on.

The scale of measurement is from 1 to 14. However, on a pH scale:

- Reading of 7 is neutral
- Reading of more than 7 is alkaline
- Reading of less than 7 is acidic

Aqueous solutions with higher $[H^+]$ than $[OH^-]$ are acidic, and will have pH values <7; basic solutions have pH >7, i.e. $[H^+] < 1x10^{-7}$.

The pH of rainwater is about 5.5-6.0. Typically, natural water has pH 6.5-8.5. Natural alkalinity is due to the dissolution of CO_2 either of atmospheric nature or from carbonate rocks, such as limestone and dolomite.

Alkalinity is increased by caustic substances from industry (KOH, NaOH), soil additives in agriculture, such as lime $Ca(OH)_2$, superphosphate which is mixture of $Ca(H_2PO_4)_2$ and $CaSO_4$, and soaps and detergents. Natural acidity is due to CO_2, HPO_4, H_2PO_4, H_2S, Fe, other acidic metal ions, proteins and organic acids. Increases in acidity can be due to acids used in industry, acid mine drainage and acid rain.

Apparatus

a) pH meter with combined electrode
b) Beakers: preferably use polyethylene or TFE beakers
c) Mechanical stirrer, with inert plastic coating
d) Wash bottle, plastic

Reagents

A. **0.01 M Potassium chloride (KCl):** Dissolve 0.745 g KCl in distilled water, and bring to 1 litre volume.

B. **Buffer solutions for pH 4 and 7**

i) **pH 4.0**

- Dry about 15 g potassium dihydrogen phthalate ($C_8H_5O_4K$) in an oven at 110°C for 3 hours.

- Dissolve 10.12 $C_8H_5O_4K$ in freshly distilled water, and bring to 1 litre volume.

ii) **pH 7.0**

- Dry about 5 g potassium di-hydrogen phosphate (KH_2PO_4) and sodium hydrogen phosphate (Na_2HPO_4) in an oven between 110°C and 130°C overnight.
- Dissolve 3.388 g KH_2PO_4 and 3.533 g Na_2HPO_4 in freshly distilled water, and bring to 1 litre volume.

  Note : The distilled water should be freshly boiled, cooled and free of CO_2.

C. **Distilled water (DI) :** Boil and cool distilled water having conductivity less than 2 μmhos/cm.

Procedure

1. Calibrate the pH meter
2. Take a 50 ml water sample in 100 ml flask, put the **combined electrode** in the water sample (about 3 cm deep). Take the reading after 30 seconds.
3. Remove the **combined electrode** from the sample and rinse thoroughly with distilled water in a separate beaker and carefully dry excess water with a tissue.

Technical remarks

- Make sure that the combined electrode contains saturated KCl solution and some solid KCl.
- If the pH meter and combined electrodes are not to be used for extended periods of time, the instructions for storage published by the instrument manufacturer should be followed.
- However, keep electrodes wet by returning them to storage solution, generally have a conductivity >4000 ìmhos/cm, such as tap water (better substitute than distilled water), pH 4.0 buffer is best for the single-glass electrode, saturated KCl is preferred for a calomel and Ag/AgCl reference electrode, and saturated KCl is the preferred solution for a combination electrode.
- If there is adequate amount of water available for testing, discard the sample used for pH determination, as it may be affected by

slight leakage from types of saturated potassium chloride-calomel reference electrodes. If there is a limited supply of liquid for analysis, determine EC first, then pH; the same sample may subsequently be used for cation-anion analysis except K and Cl.

- ACS-grade chemicals are generally satisfactory for preparing buffer solutions.
- In preparing buffer solutions from solid salts, ensure complete dissolution.

Interpretation

S. No.	pH	Category	Interpretation
1	< 6.5	Acidic	Can be used after treatment with amendments
2	6.5 -7.5	Normal	Can be used safely
3	> 7.5	Alkaline	Can be used after treatment with amendments

Chapter 26

Estimation of Electrical Conductivity (EC) of Irrigation Water

Electrical conductivity is the measure of the ability of a solution to carry an electric current or the concentration of soluble salts in the sample at any particular temperature. The EC measurement is affected by dissolved CO_2, turbidity, temperature and the nature of various ions, their relative concentration and the ionic of water.

The EC is measured in the field or in the laboratory using a conductivity meter, which is basically a Wheatstone bridge that measures the resistance of the solution between two parallel platinum electrodes. Conductivities range from less than 0.02 dS/m for distilled water to more than 20 dS/m for highly saline water.

Apparatus

Conductivity meter,thermometer, covering room temperature, conductivity cell and pipette-type or dip-type beaker (50 or 100 ml).

Reagents

0.01N Potassium chloride solution (KCl): Dry about 2-3 g KCl in an oven at 110°C for at least 2 hours.Dissolve 0.7456 g KCl in distilled water and bring to 1 litre volume. The solution should be transferred to a plastic flask. The stability of the solution is one year at most.

Procedure

A. Pre-treatment the conductivity cell

1. Calibrate the conductivity meter according to maker's instructions
2. Rinse the **conductivity cell** thoroughly with distilled water and carefully dry excess water with a tissue.

3. Rinse the **conductivity cell** thoroughly with measured solution a few times.

B. Measurement

1. Put about 75 ml water sample in a 100 ml glass beaker and then put the clean and dried conductivity cell in the glass beaker.
2. Take the reading. The display will also need some time to stabilize before the reading.
3. Remove the conductivity cell from the glass beaker, rinse thoroughly with distilled water and carefully dry excess water with a tissue.

Technical remarks

- It is essential that the manufactures instructions be followed.
- Cleaning of the **conductivity cell** is needed, if contaminated.
- The reference temperature should be 25°C and the result expressed in dS/m. If the measurement is carried out at a different temperature, the result should be corrected to 25°C. Check accuracy of the EC meter using a **0.01NKCl solution,** which should give a reading of **1.413 dS/m at 25°C**.
- Readings are recorded in milli-mhos per centimeter (mmhos/cm) or deci-siemens per metre (dS/m). The use of the unit deci-siemens is preferred over the unit milli-mhos. Both units are equal, that is, **1 dS/m = 1 mmhos/cm**.

Interpretation

S. No.	Salinity Classes	EC (μ mhos/ cm)	Remarks
1	Low (C_1)	< 250	Can be used safely
2	Medium (C_2)	250 - 750	Can be used with moderate leaching
3	High (C_3)	750 - 2250	Cannot be used for irrigation
4	Very High (C_4)	> 2250	Cannot be used for any purposes

Chapter 27

Estimation of Total Dissolved Solids (TDS) in Water

Total dissolved solid is defined as the substances remaining after evaporation and drying of awater samples. The remaining fraction is approximately equivalent to the total content of the dissolved and suspended matter in the water sample. Non-filterable residue corresponds to the **total suspended solids** (TSS) and the filterable residue is the TDS. Determination of TDS can be used tocheck the accuracy of analysis, when relatively complete analysis has been made on a water sample. This is accomplished by comparing the value of calculated TDS with the measured value. Ion concentration in mg/L of constituents, required to calculate the TDS are as follow:

Calculated TDS= 0.6 (alkalinity) =Na+K+Ca+Mg+Cl+SO_4+SiO_3+ (NO_3-N) +F

The measured TDS concentration should be higher than the calculated value, because a significant contributor may not be included in the calculation. If the measured value is less than the calculated value, all values are suspect and probably in error. If the measured value is higher than the calculated value, the low ion sum is suspect and selected constituent should be re-analysed. The acceptable ratio is:

$$10 < \frac{\text{Measured TDS}}{\text{Calculated TDS}} < 12$$

Principle

A well-mixed measured portion of a sample is filtered through a standard glass-fiber filter and the filtrate is evaporated to dryness in a weighed dish and dried to constant weight at 180°C. The increase in dish weight represents the total dissolved solids.

Apparatus

Evaporating dishes made porcelain, platinum or high-silica glass, steam bath, desiccators provided with a desiccant containing a colour indicator for moisture concentration, analytical balance capable of weighing to 0.001 g, filtration apparatus,glass-fiber filters, suction flask and drying oven.

Procedure

A. Pre-treatment of filter and crucible

1. Put the evaporating dishes for at least 1 hour in an oven at 180°C. Cool and store in desiccators until needed. Weigh immediately before use.

2. Prepare the glass-fiber disc by placing it, wrinkled side up, in the filtration apparatus. Apply vacuum and wash the disc with 3 successive 20 ml washings of distilled water. Continue suction to remove all traces of water. Discard washings.

B. Determination

1. Put an aliquot of sample to yield between 2.5 and 200 mg dried residue.

2. Filter measured volume of well-mixed water sample through the glass-fiber filter.

3. Wash with 3 successive 10 ml volumes of distilled water, allowing complete draining between washing.

4. Continue suction for about 3 minutes after filtration is complete.

5. Transfer filtrate to a weighed evaporating dish ($\mathbf{Wt_d}$) and evaporate to dryness on a steam bath.

6. Dry for at least 1 hour in an oven at 180°C, cool in desiccators, and weigh ($\mathbf{Wt_{d+s}}$).

7. Repeat the cycle of drying, desiccating and weighing until a constant weight is obtained or until weight loss between successive weighing is less than 4 % or 0.5 mg, which is less.

Calculation

$$\text{Total dissolved solids(mg/l)} = \frac{Wt_{d+s} - Wt_d}{V} \times 100$$

Where,

Wt_{d+s}= Weight of dish plus solids (mg)

Wt_d = Weight of dish before use (mg)

V= Volume of water sample used for measurement (ml)

Chapter 28

Estimation of Total Suspended Solids (TSS) in Water

Total suspended solids represents dry weight of the material that is removed from a measured volume of water sample by filtration with a standard filter. The test is basically empirical and is not subject to usual criteria of accuracy. To achieve reproducibility and comparability of results requires close attention to procedural details, especially filter characteristics and time and temperature of drying.

Apparatus

Filter holder, desiccators provided with a desiccant containing a colour indicator for moisture concentration,analytical balance capable of weighing to 0.001 g, glass-fiber filter disc, suction flask apparatus 500 ml capacity, vacuum pump, crucible, drying oven and Buchner funnels.

Procedure

A. Pre-treatment of filter disc and crucible

1. Place a filter disc on the filter holder. Assemble filter holder in suction flask apparatus, connect to vacuum source and apply vacuum.
2. Wash the filter disc with 3 successive 20 ml portions of distilled water. Continue to apply vacuum for 2-3 minutes after the water has passed through the filter. Discard the filtrate.
3. Remove the filter paper from the membrane filter funnel or the Buchner funnel and place it on a supporting surface in drying oven.
4. Place the crucible (s) in the drying oven. The oven should be maintained at 105°C and drying should be continued for at least 1 hour.

5. Cool the filter (s) and crucible (s) in desiccators and weigh it on an analytical balance.
6. Repeat the cycle of drying, desiccating and weighing until the weight loss between two successive series operations is less than 0.5 mg.
7. Store filter (s) and crucible (s) in desiccators until required.

B. Determination

1. Remove the filter disc and crucible from the desiccators and weigh (**Wt_{f+c}**).
2. Place the filter in the filter holder and assemble the filter holder in the suction flask apparatus. Connect to the vacuum source and apply vacuum.
3. Wet the filter with a few drops of distilled water to seat the filter.
4. Shake the sample vigorously and measure out 100 ml in a 100 ml graduated cylinder or volume flask. Pour this portion of the sample into the filter funnel (be careful not to disturb the placing of the filter disc).
5. Rinse out the measuring flask or cylinder with a small quantity of distilled water. If the sample is very low in suspended material, a large volume of sample may be used.
6. When filtration is complete, carefully remove the filter disc from the filter holder with tweezers (or remove the crucible from its supporting socket with a pair of tongs), and place it in the drying oven.
7. Dry for at least 1 hour at 105 °C. Cool in desiccators and weigh (Wt_{f+c+s}).
8. Repeat the drying, desiccating and weighing cycle until the weight loss between 2 successive weighing is less than 0.5 mg.
9. Record the final weight obtained.

Calculation

$$\text{Total suspended solids(mg/l)} = \frac{Wt_{f+c+s} - Wt_{f+c}}{V} \times 100$$

Where,

Wt_{f+c+s} = Weight of filter and crucible plus solids (mg)

Wt_{f+c} = Weight of filter and crucible before use (mg)

V = Volume of water sample used for measurement (ml)

Technical remarks

- It is necessary to report the specifications followed, e.g., TSS at ...°C, type of filter and pore size or number.
- Non-homogeneous particulates, such as leaves, sticks and lumps of fecal matter should be excluded from the sample. Too much residue on the filter will retain water and may require prolonged drying.
- Long-term preservation of a sample is not practical; analysis should, therefore, begin as soon as possible. Transportation and short-term storage of sample normally does not affect the results of the test.
- Volatile material in a sample can distort results.

Chapter 29

Estimation of Nitrogen in Irrigation Water

A. Ammonium nitrogen and nitrate nitrogen

In water and waste water, the forms of N of greatest interest are NO_3-N, NO_2-N, NH_3 and organic forms of N. These N forms are interchangeable, being components of the N cycle. Ammonia is present naturally in surface water and wastewater. Its concentration is generally low in ground water because it is adsorbed by soil particles and clay and is not leached readily from soils. Nitrate in groundwater is a major environmental and public health concern. Nitrates are converted to nitrites in the intestines. Once absorbed into bloodstream, nitrites prevent haemoglobin from transporting oxygen (older children have an enzyme that re-stores haemoglobin). **High NO_3-N levels in drinking water (>10 ppm) are linked with health problems** (i.e., methemoglobinemia). Thus, while using water with high NO_3-N. For human consumption should be avoided, such water is a suitable for irrigation, as it is an additional source of nutrients.

Principle

Ammonium nitrogen and NO_3-N plus NO_2-N are determined by steam distillation, using **heavy MgO** for NH_4 and **Devarda's alloy** for NO_3. The distillate is collected in saturated H_3BO_3 and titrated to pH 5.0 with dilute H_2SO_4.

Apparatus

Distillation unit, stirrer and automatic titerator connected to a pH meter.

Reagents

1. **Magnesium oxide (MgO) powder :** Heat heavy magnesium oxide in a muffle furnace at 600-700°C for 2 hours and cool in a desiccator containing KOH pellets and store in a tightly stoppered bottle.
2. **Devarda's alloy (50 Cu: 45 Al: 5 Zn) :** Ball-mill reagent-grade Devarda's alloy until the product will pass a 100 mesh sieve (0.150 mm) and at least 75% will pass a 300 mesh sieve (0.05 mm).
3. **Boric acid solution (H_3BO_3)**
   - Add 500 g H_3BO_3 into a 5 litre volume.
   - Add 3litre distilled water, and swirl vigorously.
   - Leave overnight.

   Note: There should always be solid H_3BO_3 on the bottom of the flask.
4. **0.01 N Tris solution (hydroxymethylaminomethane) ($C_4H_{11}NO_3$) :** Dry reagent-grade Tris in an oven at 80°C for 3 hours, cool in a desiccator and store in a tightly stoppered bottle.Dissolve 1.2114 g Tris in distilled water and bring to 1 litre volume.
5. **0.01 N Sulphuric acid solution (H_2SO_4) :** Add 28 ml concentrated H_2SO_4 to about 600 - 800 ml distilled water in a 1 litre flask, mix well, let it cool, and bring to 1 litre volume. This solution contains 1 N H_2SO_4 solution (stock solution). Pipette 10 ml stock solution to 1 litre flask and bring to volume with distilled water. This solution contains 0.01 N H_2SO_4.
6. **Standard stock Solution :** Dry reagent-grade ammonium sulphate $(NH_4)_2SO_4$ and potassium nitrate (KNO_3) in an oven at 100°C for 2 hours, cool in a desiccator and store in a tightly stoppered bottle.Dissolve 5.6605 g $(NH_4)_2SO_4$ and 8.6624 g KNO_3 in distilled water and bring to 1 litre volume. This solution contains (1.2 g NH_4-N and 1.2 g NO_3-N)/L (Stock Solution).

Prepare a standard solution from the stock solution as follows :

- Dilute 50 ml stock solution to 1 litre volume by adding 2 M KCl solution (diluted stock solution).
- A 20 ml aliquot of diluted stock solution contains 1.2 mg NH_4-N and 1.2 mg NO_3-N.

Procedure

A. Pre-treatment of the distillation unit

1. The distillation unit should be steamed out for at least 10 minutes. Adjust steam rate to 7-8 ml distillate/minute.
2. Water should flow through the condenser jacket at a rate sufficient to keep distillate temperature below 22°C.

B. Distillation

1. Before starting a batch for distillation, the distillation unit should be steamed out for at least 10 minutes. Adjust steam rate to 7-8 ml distillate per minute. Water should flow through the condenser jacket at a rate sufficient to keep distillate temperature below 22°C.
2. Calibrate pH meter with buffer solutions of pH 7.0 (buffer) and 4.0 (sensitivity), after setting for temperature. Then standardize the **0.01NH_2SO_4** in the **auto-titrator** by titrating three separate 10 ml aliquots of the primary standard, **0.01NTris** solution, to pH 5.0. The titrations should agree within 0.03 ml; if not; titrate further aliquots until agreement is found. The **H_2SO_4** normality is:

$$N_{H_2SO_4} = \frac{10 \text{ X } N_{tris}}{V_{H_2SO_4}}$$

3. Carry out distillations as follows:

   To determine NH_4-N:

   1. Pipette 20 ml water or wastewater sample into a 100 ml distillation flask.
   2. Pipette 1 ml **saturated H_3BO_3** solution and 1 ml distilled water into a 50ml beaker (duplicate beakers).
   3. Place the first beaker underneath the condenser tip, with the tip touching the solution surface.
   4. Add about 0.2 g **heavy MgO** with a calibrated spoon to the distillation flask.
   5. Immediately attach the distillation flask to the distillation unit with a clamp.
   6. Start distillation and continue for 3 minutes then lower the dish to allow distillate to drain freely into the Pyrex evaporating dish or beaker.

7. After 4 minutes, when 35 ml distillate or more is collected, turn off the steam supply and remove the distillation flask (**first distillate**).
8. Each distillation should contain at least two standards (pipette 20 ml 1.2 mg NH_4-N from diluted stock solution) and two blanks (pipette 20 ml 2 KCl solution). Recovery of NH_4-N should be at least 96%.

To determine NO_3-N and NO_2-N :

1. Place the second beaker underneath the condenser tip, with the tip touching the solution surface.
2. Immediately add 0.2 g **Devarda's alloy**, with a calibrated spoon, **to the same distillation flask**, then attach back to distillation unit with a clamp, and start distilling.
3. After 4 minutes, when 35 ml distillate or more is collected, turn off the steam supply and remove the distillation flask (**second distillate**).
4. Wash tip of the condenser into Pyrex evaporating dish or the beaker with a small amount of distilled water.
5. Each distillation should contain at least two standards (pipette 20 ml 1.2 mg NO_3-N from diluted stock solution) and two blanks (pipette 20 ml distilledwater). Recovery of NO_3-N should be at least 96%.

B. Titration

Titrate the first distillate (for ammonia) and the second distillate (for nitrate), separately, to **pH 5** with standardized **0.01 N H_2SO_4** using an **auto-titrator**.

Note

- After finishing titration, wash the Teflon-coated magnetic stirring bar, the burette tip, and the combined electrode into the dish.
- Between different samples, steam out the distillations. Disconnect distillation flasks containing the water sample and attach a 100 ml empty distillation flask to distillation unit and place a 100 ml empty beaker underneath the condenser tip, turn off cooling water supply (drain the water from the condenser jacket), and steam out for 90

seconds. Steaming-out is done only between distillation for different samples, not between ammonium (MgO) and nitrate (Devarda's alloy) in the same sample.

Calculation

$$NH_4\text{-N or } NO_3\text{-N (ppm)} = \frac{(V\text{-}B) \times N \times 14.01 \times 1000}{V_1}$$

Where,

V = Volume of 0.01NH_2SO_4titrated for the water sample (ml)

V_1= Volume of water sample used for distillation (ml)

B = Distillate blank titration volume (ml)

14.01 = Atomic weight of N

N = Normality of H_2SO_4solution

Technical Remarks

- If possible, NH_4-N and NO_3-N should be determined immediately after sampling. If the analysis cannot be done immediately, water and wastewater samples may be kept refrigerated (4°C).
- All water for reagents and dilution should be NH_3 free.
- Filter paper may contain traces of NH_4, so the first 20-25 ml of filtrate should be discarded.
- Never use acid preservation for samples to be analyzed for NO_3 or NH_4.

Chapter 30

Estimation of Total Nitrogen in Irrigation Water

The Kjeldahl procedure is a good estimate of N content in the water or waste water samples. This procedure involves digestion and distillation. The water or waste water sample is digested in concentrated H_2SO_4 with a catalyst mixture to raise the boiling temperature and to promote the conversion from organic-N to ammonium-N, which is obtained by steam distillation, using excess NaOH to raise the pH. The distillate is collected in saturated H_3BO_3 and then titrated with dilute H_2SO_4 to pH 5.0 (Bremner and Mulvaney, 1982).

Apparatus

Block-digester, automatic titrator connected to a pH-meter, distillation unit and vortex tube stirrer.

Reagents

A. **Catalyst mixture (K_2SO_4 – Se), 100: 1 w/w ratio:**Grind reagent-grade chemicals separately and mix. If caked, grind the mixture in a porcelain pestle and mortar to pass a 60 mesh screen (0.250 mm), taking care not to breath Se dust or allow Se to come in contact with skin.

B. **Sulphuric acid (H_2SO_4), Concentrated (98 %, special grade 1.84)**

C. **10 N Sodium hydroxide solution (NaOH):**Dissolve 400 g NaOH in distilled water, transfer to a 1 litre heavy-walled Pyrex flask, let it cool and bring to volume.

D. **Boric Acid Solution (H_3BO_3), Saturated**

- Add 500 g H_3BO_3 to a 5 litre flask.

- Add 3 litre distilled water and swirl vigorously.
- Leave overnight.

Note: There should always be solid H_3BO_3 on the bottom of the flask.

E. **0.01 NTris solution [hydroxymethylaminomethane] ($C_4H_{11}NO_3$):** Dry reagent-grade Tris in an oven at 80°C for 3 hours, cool in a desiccator, and store in a tightly stoppered bottle.Dissolve 1.2114 g Tris in distilled water, and bring to 1 litre volume

F. **0.01 NSulphuric acid solution (H_2SO_4):** Add 28 ml concentrated H_2SO_4 to about 600 - 800 ml distilled water in a 1 litre flask, mix well, let it cool, and bring to 1 litre volume. This solution contains 1 N H_2SO_4 solution (Stock Solution). Pipette 10 ml Stock Solution to 1 litre flask, and bring to volume with distilled water. This solution contains 0.01 N H_2SO_4.

G. **Standard stock solution:** Dry reagent-grade ammonium sulphate $(NH_4)_2SO_4$ in an oven at 100°C for 2 hours, cool in a desiccator and store in a tightly stoppered bottle.Dissolve 5.6605 g dried $(NH_4)_2SO_4$ in distilled water, and bring to 1 litre volume. This solution contains 1.2 g NH_4-N/L (stock solution).

Procedure

A. Digestion

1. Pipette 20 ml aliquot of the water sample into a 100 ml calibrated digestion tube.
2. Add about 3.0-3.5 g **catalyst mixture,** a few **pumice boiling granules,** add 10 ml **concentrated H_2SO_4**(in the fume hood) and then swirl carefully. Place the tubes in the rack and put a glass funnel in the neck of the tube.
3. Place the tubes rack in the block-digester, and slowly increase temperature setting to about 370-380°C. The H_2SO_4 should condense about half-way up the tube neck; when solution clears, continue heating for about 3 hours.
4. Lift the tubes rack out of the block-digester, carefully place on a rack holder, and let tubes cool to room temperature.
5. Slowly add about 15 ml distilled water to the tubes, cool, and bring to volume with distilled water.

6. Each batch of samples for digestion should contain at least one reagent blank (distilled water sample), and one chemical standard (distilled water sample, pipette 1 ml Stock Solution).

B. Distillation

1. Before starting a batch for distillation, the distillation unit should be steamed out for at least 10 minutes. Adjust steam rate to 7-8 ml distillate per minute. Water should flow through the condenser jacket at a rate sufficient to keep distillate temperature below 22°C.

2. Calibrate pH meter with buffer solutions of pH 7.0 (buffer), and 4.0 (sensitivity), after setting for temperature. Then standardize the **0.01NH_2SO_4** in the **auto-titrator** by titrating three separate 10 ml aliquots of the primary standard, **0.01NTris** solution, to pH 5.0. The titrations should agree within 0.03 ml; if not; titrate further aliquots until agreement is found. The **H_2SO_4** normality is:

$$\text{Normality of } H_2SO_4 = \frac{10 \times N_{tris}}{V_{H_2SO_4}}$$

3. Carry out distillations as follows :

   - Dispense 1 ml **saturated H_3BO_3** solution and 1 ml distilled water into a 100 ml pyrex evaporating dish, placed underneath the condenser tip, with the tip touching the solution surface.
   - Pipette 20 ml aliquot into a 100 ml distillation flask, and add 10 ml **10NNaOH** solution.
   - Immediately attach the flask to the distillation unit with a clamp, start distillation, and continue for 3 minutes, lower the dish to allow distillate to drain freely into the dish.
   - After 4 minutes when about 35-ml distillate is collected, turn off the steam supply, and wash tip of the condenser into the evaporating dish with a small amount of distilled water.
   - Titrate the distillate to pH 5.0 with standardized **0.01NH_2SO_4** using an **auto-titrator**.

- Each distillation should contain at least two standards (pipette 10 ml digested water sample) and two blanks (pipette 10 ml digested water blank). Recovery of NH_4-N should be at least 96%.

Note

- After finishing titration, wash the Teflon-coated magnetic stirring bar, the burette tip, and the combined electrode into the dish.
- Between different samples, steam out the distillations. Disconnect distillation flasks containing the digest sample and NaOH, and attach a 100ml empty distillation flask to distillation unit. Place a 100ml empty beaker underneath the condenser tip, turn off cooling water supply (drain the water from the condenser jacket), and steam out for 90 seconds.

Calculation

$$\text{Recovery (\%)} = \frac{(V\text{-}B) \times N \times 14.01 \times 100}{V_3 \times C}$$

$$N\ (\%) = \frac{(V\text{-}B) \times N \times V_1\ 14.01 \times 100}{V_2 \times 1000}$$

Where,

V = Volume of 0.01NH_2SO_4titrated for sample (ml)

V_1= Total volume of the digest (ml)

V_2= Volume of water digest used for distillation (ml)

V_3= Volume of NH_4-N standard solution (ml)

B = Digested blank titration volume (ml)

N = Normality of H_2SO_4 solution

14.01 = Atomic weight of N

C = Concentration of NH_4-N standard solution (μg/ml).

Chapter 31

Estimation of Phosphorus in Irrigation Water

Phosphorus compounds are present in fertilizers and in many detergents. If surprising a critical level in the soil, they are carried into surface water, sewage, industrial wastes and storm runoff. In contrast to N, inorganic P forms are highly reactive and do not readily leach through soils. High concentration of P compounds may produce a secondary problem in water bodies, where algal growth is normally limited by phosphorus. In such situations, the presence of additional P compounds can stimulates algal growth and enhance eutrophication.

Phosphorus occurs in natural water and in waste water as phosphate. Groundwater rarely contains more than 0.1 mg/L phosphate, if not polluted by organic matter. Phosphorus is essential to the growth of organisms and can be the nutrient that limits biological growth in water bodies.

Principle

Phosphorus can be determined by **vanadomolybdo phosphoric acid** (colorimetric methods), where are the ammonium molybdate and potassium antimony tartrate react in acid medium with orthophosphate to form a heteropoly acid-phosphomolybdic acid that is reduced to an intensely colored molybdenum blue by ascorbic acid.

Estimation of available phosphorus

Apparatus

Spectrophotometer or colorimeter, beakers, volumetric flasks, pipettes and funnels.

Reagents

A. **5 NSulphuric acid solution (H_2SO_4):**Dilute 148 ml concentrated H_2SO_4 (in fume hood) with distilled water, mix well, let it cool, and bring to 1 litre volume.

B. ***p*-nitrophenol indicator, 0.25% *w/v***

C. **Reagent A**

- Dissolve 12 g ammonium heptamolybdate $(NH_4)_6Mo_7O_{24}$. $4H_2O$ in 250 ml distilled water **(a)**.
- Dissolve 0.2908 g antimony potassium tartrate ($KSbO.C_4H_4O_6$) in 100 ml distilled water **(b)**.
- Add both dissolved Reagents **(a)** and **(b)** to a 2 litre flask.
- Slowly add 1 litre 5 N H_2SO_4 to the mixture. Mix thoroughly, and dilute to 2 litre volume.
- Store in a dark pyrex bottle, cool place.

D. **Reagent B:**Dissolve 1.056 g L-Ascorbic acid ($C_6H_8O_6$) in 200 ml Reagent A mix well. This reagent should be prepared as required because it does not keep for more than 24 hours.

E. **Standard stock solution:**Dry about 2.5 g potassium di-hydrogen phosphate (KH_2PO_4) in an oven at 105°C for 1 hour. Cool in desiccator, and store in a tightly stoppered bottle. Dissolve 2.197 g dried KH_2PO_4 in distilled water, and bring to 1 litre volume. This solution contains 500 ppm P (stock solution).Dilute 50 ml stock solution to 250 ml volume by adding distilled water. This solution contains 100 ppm P (diluted stock solution).

Prepare a series of standard solutions from the diluted stock solution as follows:

Dilute 5, 10, 15, 20 and 25 ml diluted stock solution to 500 ml numbered flasks by adding distilled water and then bring to volume. These solutions contain 1, 2, 3, 4, and 5 ppm P, respectively.

Procedure

1. Pipette a suitable aliquot of clear filter water sample (10 ml natural water sample) into a 50 ml Erlenmeyer volumetric flask.

2. Add the required acid to all the water samples to bring the solution pH to 5.0
3. Add 8 ml **Reagent B** and dilute to 50 ml volume with distilled water, mix well.
4. Prepare a standard curve as follows:
   - Pipette 2 ml of each standard (1-5 ppm), and proceed as for the samples.
   - Also make a blank with only distilled water, and proceed as for the samples.
5. Read the absorbance of blank, standards, and samples after 10 minutes on the **Spectrophotometer** at **882 nm wavelength**.
6. Prepare a calibration curve for standards, plotting absorbance against the respective P concentrations.
7. Read P concentration in the unknown samples from the calibration curve.

Calculations

$$P\ (ppm) = ppm\ P\ (from\ calibration\ curve) \times \frac{V_1}{V}$$

Where,

V = Volume of water sample used for measurement (ml)

V_1 = Volume of flask used for measurement (ml)

Technical remarks

- The required acid amount to acidify the water sample to pH 5.0 (using 5 N H_2SO_4). This can be done by taking 10 ml distilled water and adding 2-3 drops of phenolphthalein indicator, if red colour appears, add 5N H_2SO_4 till the color disappear. Then add the required amount of acid to all the water samples (adding 1 ml 5 N H_2SO_4 is adequate to acidify each 10 ml water sample).
- If the sample solutions are too dark coloured for measurement against the highest standard, smaller water should be taken, and the calculation modified accordingly. Once the blue colour has developed, the solution cannot be diluted.

- Glassware used in P analysis should not be washed with detergents containing P (most detergents contain P).
- As glass tube density may vary, it is best to use the same tube (cuvette) for each absorbance reading on a spectrophotometer.

Interference

Arsenates (As) react with the molybdate (Mo) reagent to produce a blue colour similar to that formed with phosphate. Concentrations as low as 0.1 mg As/litre interfere with the phosphate determination. Hexavalent chromium (Cr) and NO_2 interfere to give results about 3% low at concentrations of 1 mg/litre and 10 to 15% low at 10 mg/litre. Sulphide (Na_2S) and Silicate do not interfere at concentrations of 1.0 and 10 mg/litre.

Chapter 32

Estimation of Total Phosphorus in Irrigation Water

Organically bound P and all phosphates forms are first converted to orthophosphate by an acid persulphate digestion. To release P from combination with organic matter, a digestion or wet oxidation is necessary.

Principle

The digestion sample is analysed for orthophosphate based on its reaction with a combined reagent containing ammonium molybdate, antimony potassium tartrate, and ascorbic acid to form intensely coloured **molybdenum blue**.

Apparatus

Spectrophotometer or colorimeter,block digester, digestion tube, 250 ml capacity, standard laboratory glassware: beakers, pipettes, funnels, volumetric flasks and cylinders.

Reagents

A. **5 N Sodium hydroxide solution (NaOH):**Dissolve 200 g NaOH in distilled water and transfer the solution to a 1 litre heavy- walled Pyrex flask, let it cool, and bring to volume with distilled water.

B. **5 N Sulphuric acid solution (H_2SO_4):** Dilute 140 ml concentrated H_2SO_4 solution (in fume hood) with distilled water, mix well, let it cool, and bring to 1 litre volume.

C. **Concentrated perchloric acid ($HClO_4$)**

D. **Concentrated Nitric acid (HNO_3)**

E. **1% Phenolphthalein indicator:**Dissolve 1 g phenolphthalein indicator in 100 ml ethanol.

F. **Reagent A**

- Dissolve 12 g ammonium heptamolybdate $(NH_4)_6Mo_7O_{24}.4H_2O$ in 250 ml distilled water (a).
- Dissolve 0.2908 g antimony potassium tartrate $(KSbO.C_4H_4O_6)$ in 100 ml distilled water (b).
- Add both dissolved Reagents (a) and (b) to a 2 litre flask.
- Slowly add 1 litre 5 N H_2SO_4 to the mixture. Mix thoroughly and dilute to 2 litre volume.
- Store in a dark Pyrex bottle, cool place.

G. **Reagent B:**Dissolve 1.056 g L-Ascorbic acid $(C_6H_8O_6)$ in 200 ml Reagent A, mix well. This reagent should be prepared as required because it does not keep for more than 24 hours.

H. **Standard stock solution:**Dry about 2.5 g potassium dihydrogen phosphate (KH_2PO_4) in an oven at 105 °C for 1 hour, cool in a desiccator, and store in a tightly stoppered bottle.Dissolve 2.197 g dried KH_2PO_4 in distilled water, and bring to 1 litre volume with distilled water. This solution contains 500 ppm P (stock solution). Dilute 50 ml Stock Solution to 250 ml volume by adding distilled water. This solution contains 100 ppm P (diluted stock solution).

Prepare a series of standard solutions from the diluted stock solution as follows:

Dilute 5, 10, 15, 20 and 25 ml diluted stock solution to 500 ml numbered flasks by adding distilled water and then bring to volume. These solutions contain 1, 2, 3, 4 and 5 ppm P, respectively.

Procedure

A. Digestion

1. Add a suitable aliquot of a well-mixed water sample (10 ml natural water sample), into a 250 ml calibrated digestion tube. Rinse the cylinder twice with distilled water.
2. Add 5 ml **concentration HNO_3** and a few **pumice-boiling granules**, mix well, and then place tubes in the rack.

3. Place the tubes rack in the block digester and gently heat to evaporate to about 15-20 ml (it takes 1-1.5 hour).
4. Cool, and then add 10 ml **(2:1 ratio) concentration HNO_3** and **$HClO_4$ mixture.**
5. Again, place the tubes rack in the block digester and slowly increase the block digester temperature to 180°C and digest the samples until dense white fumes of **$HClO_4$** appear (it takes 1.5-2 hours).
6. If solution is not clear, continue heating at the boiling temperature for 15-20 minutes longer. If necessary, add 10 ml more **concentration HNO_3** to aid oxidation.
7. Cool the mixture and add small amount of distilled water, mix well, and then bring to 50 ml volume with distilled water. If necessary, filter through Whatman No. 1 filter paper.

B. Measurement

1. Pipette 5-10 ml of the sample digest into a 50 ml flask.
2. Add 0.5 ml **5MNaOH** and 1 drop **phenolphthalein** solution to each sample and mix well.
3. Add 0.4 ml **6MNaOH** until the solution just turns pink, and then add **$6MH_2SO_4$** until the pink color just clears.
4. Add distilled water to about 40 ml volume, add 8 ml **Reagent B,** and then bring to volume.

Prepare a standard curve as follows :

A. Pipette a suitable volume of each standard (0.2-2 ppm), and proceed as for the samples.

B. Also, make a blank, and proceed as for the samples.

C. Read the absorbance of blank, standards, and samples after 10 minutes on the **Spectrophotometer** at **882 nm wavelength**.

D. Prepare a calibration curve for standards, plotting absorbance against the respective P concentrations.

E. Read P concentration in the unknown samples from the calibration curve.

Calculations

$$P\text{ (ppm)} = \text{ppm P (from calibration curve)} \times \frac{V_1}{V} \times \frac{V_3}{V_2}$$

Where,

V = Total volume of the water sample used for digestion (ml)

V_1 = Total volume of the digest (ml)

V_2 = Volume of digest used for measurement (ml)

V_3 = Volume of flask used for measurement (ml)

Caution: $HClO_4$ is explosive in presence of easily oxidizable organic matter. Avoid this hazard by taking the following precaution: (a) do not add $HClO_4$ to a hot solution containing high organic matter; (b) always pre-treat samples containing organic matter with HNO_3 before adding $HClO_4$.

Technical remarks

- No interferences are normally observed for Cu, Fe, or Si. However, high concentrations of Fe can cause precipitation and subsequent loss of P.
- Arsenate may interfere, when present at concentration higher than P.

Phosphate adsorbed on glass surface may affect measurements at low P levels. Use of acid-washed glassware dedicated to this analysis prevents this.

Chapter 33

Estimation of Potassium in Irrigation Water

Although **potassium (K)** is a relatively abundant element, its concentration in natural fresh water is usually less than 20 mg/L. Saline and seawater, however, may contain as much as 400 mg/L of K or more. Potassium in water can be determined by **flame photometry**.

Principle

The estimation of K is based on the emission spectroscopy, which deals with excitation of electrons from ground state to a higher energy state and coming back to its original state with the emission of light.

Apparatus

Flame photometer with accessories, beakers, pipettes and volumetric flasks are required for dilution and tests of interference effects.

Reagents

Standard Stock Solution: Dry 3-5 g potassium chloride (KCl) in an oven at 120°C for 1-2 hours and cool in a desiccator, and store in a tightly stoppered bottle. Dissolve 1.907 g dried KCl in distilled water and bring to 1 litre volume. This solution contains 1000 ppm K (stock solution).

Prepare a series of standard solutions from the stock solution as follows:

Dilute 2, 4, 6, 8, 10, 15 and 20 ml stock solution to 100 ml numbered flasks by adding distilled water and then bring to volume. These solutions contain 20, 40, 60, 80, 100, 150, and 200 ppm K, respectively.

Procedure

1. Filter a portion of water sample through Whatman no. 42 filter paper.
2. For operating the **flame photometer**, follow the maker's instructions.
3. Calibrate **flame photometer** with a series of suitable K standards with distilled water as blank sample.
4. Measure the water samples, take the emission reading on the **flame photometer** at **767 nm wavelength**, and record the readings.
5. Switch off the flame photometer according to the maker's instructions.
6. Draw a calibration curve.
7. Calculate K concentrations according to the calibration curve.

Calculations

K (ppm) = ppm K (from calibration curve)

K (ppm) = ppm K (from calibration curve)/39.1

Where,

39.1 = Atomic weight of K

Technical remarks

- Check the performance of the photometer at frequent intervals by spraying some of the standard solutions and adjust the sensitivity as necessary.
- If K concentration is higher than the top standard, make an appropriate dilution. If this is a dilution of the original sample, multiply by the appropriate factor.

Chapter 34

Estimation of Sodium in Irrigation Water

Sodium (Na) is a common element, the sixth most abundant and present to some extent in most natural water. Sodium is present in a number of minerals, the principal one being rock salt (sodium chloride). The increased pollution of surface and groundwater during the past decade has resulted in a substantial in the Na content of drinking water in different regions of the world. Sewage, industrial effluents, seawater intrusion in coastal area, and the use of Na compounds for corrosion control and water softening processes all contribute to Na concentration in water because of the high solubility of sodium salts and minerals.

Sodium levels in groundwater vary widely but normally range between 6 and 130 mg/L. In surface water the Na concentration may be less than 1 mg/L or exceed 300 mg/L depending upon the geographical area. Sodium in water exceeds 20 mg/L becomes a problem.

Principle

Sodium in the water sample can be determined by flame photometry emitting a sparkling yellowish-red colour.

Reagents

A. **100 ppm Lithium chloride (LiCl):**Dissolve 6.109 g dry LiCl in distilled water and bring to 1 litre volume. This solution contains 1000 ppm LiCl (stock solution).Dilute 100 ml stock solution to 1 litre volume. This solution contains 100 ppm LiCl (diluted stock solution).

B. **Standard stock solution:**Dry 5 g sodium chloride (NaCl) in an oven at 110°C for 3 hours, cools in a desiccator, and store in a tightly stoppered bottle. Dissolve 2.5418 g dried NaCl in distilled water, and bring to 1 litre volume. This solution contains 1000 ppm Na (stock solution).

Prepare a series of standard solutions from the stock solution as follows:

Dilute 2, 4, 6, 8, 10, 15, and 20 ml stock solution to 100 ml numbered flask by adding distilled water, and 25 ml LiCl diluted stock solution, and then bring to volume. These solutions contain 20, 40, 60, 80, 100, 150, and 200 ppm Na, with each containing the same concentration of LiCl (25 ppm).

Procedure

1. Filter a portion of water sample through Whatman filter paper No. 42.
2. For operating the **flame photometer**, follow the maker's instructions.
3. Calibrate **flame photometer** with a series of suitable Na standards with distilled water as blank sample.
4. Measure the water samples, take the emission reading on the **flame photometer** at **589 nm wavelength**, and record the readings.
5. Switch off the flame photometer according to the maker's instructions.
6. Draw a calibration curve.
7. Calculate Na concentrations according to the calibration curve.

Calculations

Na (ppm) = ppm Na (from calibration curve)

Na (meq/l) = ppm Na (from calibration curve)/23

Where,

23 = Atomic weight of Na

Technical remarks

- In the analysis of natural water, the only ions which may cause serious interference to Na measurements is calcium; in some types of flames, this tends to enhance the Na emission and the effect may be measured for a range of Ca concentrations.

- Sodium ratios and appropriate corrections applied or the interference may be neutralized by the addition of aluminium (Al) ion, usually as aluminium nitrate.

Chapter 35

Estimation of Calcium and Magnesium in Irrigation Water

Calcium (Ca) is dissolved easily out of almost all rocks and is consequently detected in most of the water. **Magnesium** (Mg) arerelatively abundant in the earth's crust and hence, a common constituent of natural water. Water associated with granite or siliceous sand usually contain less than 10 mg of calcium per litre and less than 5 mg magnesium per litre.

Many kinds of water from limestone areas may contain 30-100 Ca/L and those associated with gypsiferous shale may contain several hundred milligrams per litre. But for the water in contact with dolomite or Mg-rich limestone may content 10-50 mg/L and several hundred milligrams per litre may be present in water that has been in contact with deposits containing sulphates and chlorides of magnesium.

Calcium and Mg contribute to the total hardness of water. It should be noted that the difference between total hardness and the Ca concentration can be used to calculate the magnesium concentration. However, some $CaCO_3$ is desirable for domestic water because it provides a coating in the pipes which protects them against corrosion.

Principle

EDTA-disodium salt solution is used to chelate Ca + Mg. Calcium is separately estimated by the versenate method using ammonium purpuate (Murexide) indicator, when the pH is made sufficiently high that the Mg is largely precipitated as hydroxide and an indicator is used that combines with Ca only. Thus, Mg can be obtained by deduction of Ca from Ca+Mg content. Both cations can also estimate by atomic absorption spectrophotometer.

Apparatus

Burette (25 or 50 ml), graduated cylinder, pipette, beakers and stirring rods.

Reagents

A. **Buffer solution (NH_4Cl-NH_4OH):** Dissolve 67.5 g NH_4Cl in 570 ml concentrated NH_4OH, and transfer the solution to a 1liter flask, let it cool, and bring to volume.

B. **Eriochrome black indicator:** Dissolve 0.5 g eriochrome black and 4.5 g hydroxylamine hydrochloride in 100 ml 95% ethyl alcohol. Prepare a fresh batch every month.

C. **0.01 N Ethylene di-aminetetraacetic acid solution (EDTA):** Dissolve 2 g EDTA, and 0.05 g magnesium chloride ($MgCl_2$) in distilled water, and bring to 1 litre volume.

D. **2 N Sodium hydroxide solution (NaOH):** Dissolve 80 g NaOH in about 800 ml distilled water, transfer the solution to a 1 litre flask, cool, and bring to volume.

E. **Ammonium purpurate indicator ($C_8H_8N_6O_6$):** Mix 0.5 g ammonium purpurate (Murexide) with 100 g potassium sulphate (K_2SO_4).

F. **0.01 N Standard stock calcium chloride solution ($CaCl_2.2H_2O$):** Dry about 3 g $CaCO_3$ in an oven at 100°C for 3 hours, cool in a desiccator, and store in a tightly stoppered bottle.Dissolve 0.5 g dried $CaCO_3$ in 10 ml 3 N hydrochloric acid (HCl) and bring to 1 litre volume with distilled water.

G. **AN alternative preparation**: Standard stock can also be prepared by dissolving 0.735 g calcium chloride di-hydrate ($CaCl_2.2H_2O$) in 1 liter volume with distilled water.

Procedure

A. Calcium

1. Pipette a suitable aliquot of water sample (10 ml natural water sample) into a 250 ml Erlenmeyer flask.
2. Dilute to 20-30 ml with distilled water, add 2-3 ml **2NNaOH** solution, and about 50 mg **ammonium purpurate** indicator.

3. Titrate with **0.01 NEDTA**. The color change is from red to lavender or purple. Near the endpoint, **EDTA** should be added one drop every 10 seconds since the colour change is not instantaneous.

B. Calcium +Magnesium

1. Pipette a suitable aliquot of water sample (10 ml natural water sample), dilute to 20-30 ml with distilled water. Then add 3-5 ml **buffer solution**. And a few drops **eriochrome black** indicator.
2. Titrate with **0.01NEDTA** until the colour changes from red to blue.
3. In order to standardize the EDTA solution used in the determination of Ca and Mg:
   - Pipette 10 ml 0.01 N calcium chloride solution, and treat it as in determining Ca and Ca+Mg procedure, respectively.
   - Take the reading, and calculate **EDTA** normality:

$$N_{EDTA} = \frac{10 \text{ x } N_{Cacl_2}}{V_{EDTA}}$$

Where,

N_{EDTA} = Normality of EDTA solution

V_{EDTA} = Volume of EDTA solution used (ml)

N_{CaCl2}= Normality of $CaCl_2$ solution

Calculations

$$\text{Ca or Ca + Mg (meq/l)} = \frac{V_1 \text{ X N X } 1000}{V}$$

Mg (meq/l) = [Ca + Mg (meq/l)] -Ca(meq/l

Where,

V_1 = Volume of EDTA titrated for the sample (ml)

N = Normality of EDTA solution

V = Volume of water sample used for measurement (ml)

Technical remarks

- Normality with Ca determination usually is 3 to 5% higher than with Ca + Mg.

- If an Atomic Absorption Spectrophotometer is used, a small aliquot of the water sample is sufficient to determine Ca and Mg.
- Orthophosphate precipitates Ca at pH of the test. Strontium (Sr) and barium (Ba) interfere with the Ca determination and alkalinity in excess of 300 mg/L may cause an indistinct end-point with hard water. Under the conditions of the test, normal concentration of the following ions causes no interference with the Ca determination: Cu, Fe, Mn, Zn, Al, Pb, Cu, and Sn.
- **Calcium:** Low Ca intake can be related to hypertension and cardiovascular disorders. There is a link between low Ca intake and osteoporosis. With a low level of Ca in the diet, drinking water may provide a significant portion of the daily Ca requirement.
- **Magnesium:** An average adult ingests as much as 480 mg of Mg daily. Any excess amounts are quickly expelled by the body. No upper limit has been set for this metal in drinking water. It can, however, create a problem for people with kidney disease. They may develop toxic reactions to high levels of Mg, including muscle weakness, coma, hypertension, and confusion.
- **Hardness:** In most water nearly all of the hardness is due to Ca and Mg, which react withsoap to form precipitates.This increases soap consumption, and react with certain constituents to form scale. As a general rule, a value less than 60 is considered soft, and values above 200 are considered very hard.

Chapter 36

Estimation of Carbonates and Bicarbonates in Irrigation Water

Carbonates (CO_3) and **Bicarbonates** (HCO_3) in water can be determined by titration a known volume of water against standard H_2SO_4 using **phenolphthalein** and **methyl orange** indicators, respectively.

Principle

Carbonates

When the pH value of a sample of natural water is above 8.4, the CO_3 is present, normally as sodium carbonate. The carbonate ion is converted to HCO_3. The amount of acid used is a measure of the carbonate present.

$$CO_3 + H = HCO_3$$

Bicarbonates

Bicarbonate ions react with mineral acid and release carbon dioxide (CO_2) into the solution.

$$HCO_3 + H = H_2O + CO_2$$

The pH value at complete neutralization being about 3.8, HCO_3 ions are present. Thus, HCO_3 may be measuring by titration with mineral acid to a pH 3.8, either potentiometrically or using an indicator unaffected by CO_2. Methyl orange is suitable and gives a good colour change from green through grey (endpoint) to red, which avoids the need for a matching buffer solution.

Apparatus

Pipettes, graduated pipette, burette, magnetic stirrer and Erlenmeyer flasks.

Reagents

A. **Methyl orange indicator:** Dissolve 0.1 g methyl orange indicator in 100 ml volume with distilledwater.

B. **0.01 N Sulphuric acid solution (H_2SO_4):** Add 28 ml concentrated H_2SO_4 to about 600-800 ml distilled water in a 1 litre flask, mix well, let it cool, and bring to 1 litre volume. This solution contains 1 N H_2SO_4 solution (stock solution). Pipette 10 ml stock solution to 1 litre flask and bring to volume with distilled water. This solution contains 0.01 N H_2SO_4.

C. **1% Phenolphthalein indicator:** Dissolve 1 g phenolphthalein indicator in 100 ml ethanol.

Procedure

1. Pipette a suitable aliquot of water sample (10 ml natural water sample) in a wide-mouthed porcelain crucible or a 250 ml Erlenmeyer flask.
2. Add 1 drop **phenolphthalein** indicator. If pink colour develops, add **0.01NH_2SO_4** by a burette, drop by drop, until the colour disappears.
3. Take the reading y.
4. Continue the titration with 0.01 N H_2SO_4 after adding 2 drops 0.1% methyl orange indicator until the colour turns to orange.
5. Take the reading t.

Calculations

$$CO_3\ (\text{meq/l}) = \frac{2y \times N \times 1000}{V}$$

$$HCO_3\ (\text{meq/l}) = \frac{(t-2y) \times N \times 1000}{V}$$

Where,

2 = Valance of carbonate

Y = Volume of titrant against phenolphthalein indicator (ml)

T = Volume of titrant against methyl orange indicator (ml)

V = Volume of water sample used for measurement (ml)

N= Normality of H_2SO_4 solution

Technical remarks

- Standard HCl is used because H_2SO_4 may give rise to turbidity from calcium sulphate with Carich samples.
- **The alkalinity of water** is a measure of its capacity to neutralize acids and is due primarily to the presence of bicarbonates. The acceptable alkalinity for municipal water supplies is generally between 30 and 500 mg/L as $CaCO_3$, but there are many water supplies above and below these limitations. Water with alkalinity greater than 500 mg/L as $CaCO_3$ have objectionable tastes.

Chapter 37

Estimation of Chloride in Irrigation Water

Chloride (Cl) anions are usually present in natural water. A high Cl concentration occurs in water that have been in contact with Cl containing geological formations. Otherwise, high Cl content may indicate pollution by sewage or industrial wastes or by intrusion of seawater or saline water into a fresh water body or aquifer. **A salty taste in water dependents on the ions with which the Cl are associated.** With Na ions the taste is detectable at about 250 mg/L, but with Ca or Mg the taste maybe undetectable at 1,000 mg/L. Chlorides being highly soluble is present in all types of water but the amount is often very low in natural water. **High Cl content has a corrosive effect on metal pipes and structures and is harmful to most trees and plants.**

Principle

The determination of Cl is done by $AgNO_3$(Mohrs's titration) method, which is based upon the factthat in solution containing Cl and chromate (Cr). Silver reacts with all the Cl and precipitates before the reaction with chromate begins. The appearance of the brick-red colour of the silver chromate precipitate is the endpoint of the titration.

Apparatus

Beakers, volumetric flask, burette and stand.

Reagents

A. **Potassium chromate solution (K_2CrO_4), 5% in water**: Dissolve 5 g K_2CrO_4 in 50 ml distilled water.Add drop-wise 1 N silver nitrate ($AgNO_3$) until a slight permanent red precipitate is formed.Filter and bring to 100 ml volume with distilled water.

B. **0.01 N Silver nitrate solution ($AgNO_3$):** Dry about 3 g $AgNO_3$ in an oven at 105°C for 2 hours, cool in a desiccator, and store in a tightly stoppered and brown bottle.Dissolve 1.696 g dried $AgNO_3$ in distilled water, and bring to 1 litre volume.

C. **0.01 N Sodium chloride solution (NaCl):** Dry about 3 g NaCl in an oven at 140 °C for 2 hours, cool in a desiccator, and store in a tightly stoppered and brown bottle.Dissolve 0.585 g NaCl in distilled water, and bring to 1 litre volume.

Procedure

1. Pipette a suitable aliquot of water sample (10 ml natural water sample) into a 250 ml Erlenmeyer flask.
2. Add 4 drops **potassium chromate** solution.
3. Titrate against **$AgNO_3$** solution until a permanent reddish-brown colour appears.
4. In order to standardize the $AgNO_3$ solution used in the determination of Cl:
   - Titrate 10 ml **0.01NNaCl** solution against **0.01N$AgNO_3$** after adding 4 drops **potassium chromate** solution until a permanent reddish-brown colour appears.
   - Take the reading, and calculate **$AgNO_3$** normality:

   $$N_{AgNO3} = \frac{10 \text{ X } N_{NaCl}}{V_{AgNO3}}$$

   Where,

   N_{AgNO3} = Normality of $AgNO_3$ solution
   V_{AgNO3} = Volume of $AgNO_3$ solution used (ml)
   N_{NaCl} = Normality of NaCl solution

Calculations

$$Cl\ (meq/l) = \frac{V_1 \text{ X } N \text{ X } 1000}{V}$$

Where,

V_1 = Volume of 0.01 N $AgNO_3$ titrated for the sample (ml)
N = Normality of $AgNO_3$ solution
V = Volume of water sample used for measurement (ml)

Technical remarks

- Natural wateris often low in Cl, and 10 ml is a suitable aliquot in most cases.
- Saline water may be high in Cl and 5 ml (or even less than 5 ml) may then be more appropriate aliquots.
- The Cl estimation has two different purposes:
  - If the test is done regularly on a water supply and there is a sudden increase, it may indicate pollution due to increase of sewage or other chloride-rich wastes.
  - Many groundwater have Cl content high enough to be of objectionable taste. By using the Cl test, the well with the lowest amount of Cl can be identified. If several well are being pumped, it can be planned in such a way that the lowest Cl content is obtained.

Interpretation

S. No.	Cl (ppm)	Classes	Water quality	Remarks
1	< 2	Very Low	Excellent water	Non saline water
2	2-5	Low	Moderately good water	Weakly saline
3	5-12	Medium	Slightly usable for irrigation	Moderately saline
4	12-20	High	Not suitable for irrigation	Strongly Saline
5	> 20	Very high	Not suitable for irrigation	Very strongly saline

Chapter 38

Estimation of Sulphate in Irrigation Water

Sulphate (SO_4)is an abundant ion in the earth's crust and its concentration in water can range from afew milligrams to several thousand milligrams/L. Industrial wastes and mine drainage may contain high concentration of SO_4, which also results from breakdown of SO_4containing organic compounds. The SO_4 is not a notably toxic anion; however, SO_4 in water containing Ca forms hard scale in steam boilers. In large amounts, SO_4 in combination with other constituents gives a bitter taste to water. Concentrations above 250 mg/L may have a laxative effect, but 500 mg/L is considered safe. **The SO_4in water is determined normally by barium sulfate ($BaSO_4$) precipitation**.

Apparatus

Muffle furnace and porcelain crucible.

Reagents

A. **Methyl orange indicator:** Dissolve 0.1 g methyl orange indicator in 100 ml distilled water

B. **1:1 Hydrochloric acid solution (HCl):** Mix equal portions of concentrated HCl with distilled water.

C. **1 N Barium chloride solution ($BaCl_2.2H_2O$):**Dissolve 122 g $BaCl_2.2H_2O$ in distilled water, and bring to 1 litre volume.

Procedure

1. Pipette a suitable aliquot of water sample in a 250 ml Erlenmeyer flask or a wide-mouthed pyrex beaker (50 ml natural water sample)

2. Add 1 ml **1: 1 HCl** solution and 2-3 drops **methyl orange**; if the colour does not turn pink; add some more 1:1 HCl.
3. Put pyrex beakers on a hotplate, heat to boiling, and then add 10 ml **1N$BaCl_2.2H_2O$** solution in excess to precipitate SO_4-S as barium sulphate ($BaSO_4$).
4. Boil for 5 to 10 minutes, cover with a watch-glass, and leave to cool.
5. Filter solution through ash less filter paper, collect the $BaSO_4$ precipitate on the filter paper, and then wash it several times with warm distilled water until no trace of Cl remains. The presence of Cl in the filtrate can be checked by $AgNO_3$ solution.
6. After washing, place filter paper with precipitate into a pre-weighed and dried porcelain crucible (**Wt_1**), and put in an oven at 105 °C for 1 hour to dry.
7. Transfer crucible to a muffle furnace heated to 550 °C, and leave to dry ash for 2 - 3 hours.
8. Take crucible out of the muffle furnace, and place in a desiccator to cool, weigh crucible on an analytical balance, and take the reading (**Wt_2**).

Calculations

$$SO_4\text{-S (meq/l)} = \frac{Wt_2 - Wt_1}{V} \times \frac{1000}{0.1165}$$

Where,

Wt_2 = Weight of crucible + $BaSO_4$ precipitate (g)

Wt_1 = Weight of empty crucible (g)

V = Volume of water sample used for measurement (ml)

Note: 0.1165 g $BaSO_4$ equal to 1 milli-equivalent of SO_4

Chapter 39

Estimation of Boron in Irrigation Water

In most natural resources, **Boron** (B) is rarely found in concentrations greater than 1 mg/L, but even this low concentration can have deleterious effects on certain agricultural products. Water having B concentrations in excess of 2 mg/L can adversely affect many common crops. However, where levels are greater than 5 mg/L, toxicity may occur. Groundwater may have a greater B concentration, particularly in areas, where the water comes in contact with igneous rocks or other B-containing strata.

The B content in many waters has been increasing as a result of the introduction of industrial waste and of the use of boric acid (H_3BO_3) and its salts in cleaning compounds. Ingestion of B at concentrations usually found in natural water has no adverse effects on humans. Ingestion of large quantities of B, however, can affect the central nervous system, while extended consumption of water containing B can lead to a condition known as borism.

The **hot-water procedure** is still the most popular method for measuring B, and it was introduced by Berger and Truog (1939), and was modified by later researchers. The B is measured colorimetrically using **Azomethine-H** (Bingham, 1982).

Apparatus

Erlenmeyer flasks, polypropylene test tubes, spectrophotometer or colorimeter.

Reagents

A. **Buffer solution:**Dissolve 250 g ammonium acetate (NH_4OAC), and 15 g EDTA disodium (ethylenediaminetetraacetic acid, disodium salt) in 400 ml distilled water. Slowly add 125 ml glacial acetic acid(CH_3COOH), and mix well.

B. **Azomethine-H Solution ($C_{17}H_{12}NNa\ O_8S_2$):**Dissolve1 g L-ascorbic acid in 100 ml distilled water, and then add 0.45 g Azomethine-H, and mix well.Fresh reagent should be prepared weekly and stored in a refrigerator.

C. **Standard stock solution:**Dissolve 0.114 g boric acid (H_3BO_3) in distilled water, and bring to 1 litre volume. This solution contains 20 ppm B (stock solution).

Prepare a series of standard solutions from the stock solution as follows:

Dilute 2.5, 5.0, 7.5, 10.0, 12.5 and 15.0 ml stock solution to 100 ml numbered flasks by adding distilled water, and then bring to volume. These solutions contain 0.5, 1.0, 1.5, 2.0, 2.5, and 3.0 ppm, respectively.

Procedure

1. Pipette 2 ml aliquot of the natural water sample into a 10 ml polypropylene tube.
2. Add 4 ml **buffer solution**.
3. Add 4 ml **Azomethine-H** solution, and mix well.
4. Prepare a standard curve as follows:
   - Pipette 2 ml of each standard (0.5-3.0 ppm), and proceed as for the samples.
   - Also make a blank with 2 ml distilled water, and proceed as for the samples.
5. Read the absorbance of blank, standards, and samples after 30 minutes on the **Spectrophotometer** at **420 nm wavelength**.
6. Prepare a calibration curve for standards, plotting absorbance against the respective B concentrations.
7. Read B concentration in the unknown samples from the calibration curve.

Calculations

B (ppm) = ppm B (from calibration curve)

Technical remarks

- Use of glassware should be minimal; always use concentrated HCl treated glassware (soaking for a week), where absolutely essential.

Interpretation

S. No.	Boron Classes	Boron (ppm) Sensitive	Moderate	Tolerant	Remarks
1	Very low	< 0.33	< 0.67	< 1.00	Can be used safely
2	Low	0.33-0.67	0.67-1.33	1.00-2.00	Can be used with management
3	Medium	0.67-1.00	1.33-2.00	2.00-3.00	Unsuitable for irrigation purposes
4	High	1.00-1.25	2.00-2.50	3.00-3.75	Unsuitable for irrigation purposes
5	Very High	> 1.25	> 2.50	> 3.75	Unsuitable for irrigation purposes

Chapter 40

Computation of SAR and RSC of Irrigation Water

The sodium adsorption ratio (SAR) and residual sodium carbonate (RSC) of irrigation water can be calculated after determining the exchangeable cations and anions of irrigation water.

1. SAR

$$SAR = \frac{Na^+}{\sqrt{Ca+Mg/2}}$$

Sodicity classes

S. No.	Sodicity Classes	SAR	Remarks
1	Low (S_1)	<10	Little or no hazard
2	Medium (S_2)	10-18	Can be used with appropriate management
3	High (S_3)	18-26	Unsatisfactory for most of the crops
4	Very High (S_4)	> 26	Unsatisfactory for most of the crops

2. RSC (meq/ 100 g) = [(CO_3+ HCO_3) – (Ca + Mg)]

Bicarbonate classes

S. No.	HCO_3 Classes	RSC (meq/L)	Remarks
1	Low (RSC_1)	<1.25	Water can be used safely
2	Medium(RSC_2)	1.25-2.50	Can be used with management
3	High (RSC_3)	> 2.50	Unsuitable for irrigation purposes

Chapter 41

Essential Nutrients and Its Functions

Three elements, carbon (C), hydrogen (H) and oxygen (O), are supplied by air (in the form of carbon dioxide) and water. When the chlorophyll (green pigments) of plants are exposed to light, these three elements are combined in a process called photosynthesis to make carbohydrates, with a subsequent release of oxygen. Then water is brought into the plant by root absorption from the soil system. Carbon dioxide (CO_2) enters the plant through small leaf openings called stomata. The rate at which photosynthesis occurs is directly influenced by the water and nutritional status of the plant. Maximum rates are determined ultimately by the genetics of the plant. Of these, nitrogen (N), phosphorus (P) and potassium (K) are referred to as primary or macro-nutrients. This is because (1) they are required by the plant in large amounts relative to other nutrients and (2) they are the nutrients most likely to be found limiting plant growth and development in soil systems.

Essential plant nutrients and their elemental (chemical) symbol

Nutrients Supplied	**Nutrients supplied by the soil system**		
Non-mineral	**Primary ormacro-nutrients**	**Secondary nutrients**	**Micronutrients**
Carbon - C Hydrogen - H Oxygen - O	Nitrogen - N Phosphorus - P Potassium - K	Calcium - Ca Magnesium - Mg Sulphur - S	Zinc - Zn Chlorine - Cl Boron - B Molybdenum - Mo Copper - Cu Iron - Fe Manganese - Mn Nickel - Ni

Calcium (Ca), magnesium (Mg) and sulphur (S) are termed secondary nutrients because they are less likely to be growth-limiting factors in soil systems. Calcium and magnesium are added in liming materials, when soil pH is adjusted and sulphur is added continually by rainfall and release from the soil organic matter. It is estimated that some 10 to 20 pounds of sulphur per acre may be deposited annually in precipitation.

Zinc (Zn), chlorine (Cl), boron (B), molybdenum (Mo), copper (Cu), iron (Fe), manganese (Mn) and nickel (Ni) are termed micro-nutrients because (1) they are found in only very small amounts relative to other plant nutrients in the average plant and (2) they are least likely to be limiting plant growth and development in many soil systems.

Functions and available forms of nutrients

Nutrient	**Functions in plants**	**Plant available from soil solution complex**	
		Form(s)	**Symbol (s)**
Nitrogen	Promotes rapid growth, chlorophyll formation and protein synthesis	Anion Cation	NO^{-3} NH^{+4}
Phosphorus	Stimulates early root growth. Hastens maturity Stimulates blooming and aids seed formation	Anion	$H\ PO^{-}HPO_4^{-4}$
Potassium	Increases resistance to drought and disease	Cation	K^+
Calcium	Improves root formation, stiffness of straw and vigor. Increases resistance to seedling diseases	Cation	Ca^{++}
Magnesium	Aids chlorophyll formation and phosphorus	Cation	Mg^{++}
Sulphur	Amino acids, vitamins. Imparts dark green colour Stimulates seed production	Anion	SO_4^{-2}
Boron	Aids carbohydrate transport and cell division	Anion	H_3BO_3
Copper	Enzymes, light reactions	Cation*	Cu^{+2}
Iron	Chlorophyll formation	Cation*	Fe^{+2}, Fe^{+3}
Manganese	Oxidation-reduction reactions Hastens germination and maturation	Cation*	Mn^{+2}
Zinc	Auxins, enzymes	Cation*	Zn^{+2}

Molybdenum	Aids nitrogen fixation and nitrate assimilation	Anion	MoO^{-2}
Cobalt	Essential for nitrogen fixation	Cation	Co^{++}
Nickel	Grain filling, seed viability	Cation	Ni^{+2}, Ni^{+3}
Chlorine	Water use	Anion	Cl^{-}
Oxygen	Component of most plant compounds	Obtained from air and water	
Hydrogen	Component of most plant compounds		
Carbon	Component of most plant compounds		

*Also available to plants in chelate form (a nutrient form having the essential nutrient linked to an organic compound so that it stays available for plant use within certain ranges of soil pH).

Critical limits of major nutrients in plants

Nutrients	Plant (%)
N	1.5
P	< 0.2
K	1-3
S	< 0.15-0.2
Ca	<0.2
Mg	0.1-0.2

Critical limits of minor nutrients in plants

Nutrients	Plant(ppm)
Zn	< 20
Fe	< 50
Mn	< 20
Cu	< 4
B	< 1
Mo	< 0.1
Cl	< 100

Chapter 42

Visual Diagnosis of Plant Nutrient Deficiencies

Sometimes, the soil chemistry is such that the soil is not able to supply sufficient nutrients to the plant. Toxic conditions, such as excessive soil acidity may prevent plant roots from growing or perhaps nutrients are simply in low supply. When these conditions are severe enough, plants will exhibit nutrient deficiency symptoms. The symptoms expressed by the plant are often used to somewhat subjectively diagnose plant nutrient problems. Some common symptoms shown by plants are:

A. **Chlorosis:** A yellowing, either uniform or interveinal of plant tissue due toa reduction of the chlorophyll formation processes.

B. **Necrosis:** The death or dying of plant tissue. It usually begins on the tipsand edges of older leaves and also may be caused by drought, herbicides, disease, and foliar application of fertilizer or animals marking territorial boundaries.

C. **Rosetting:** A cluster of leaves crowded and arising from a crown, resulting from a lack of new terminal growth.

D. **Anthocyanin (pigment) accumulation:** This results in the appearance of reddish, purple or brownish colouration. The pigment anthocyanin forms due to sugar accumulation.

E. **Stunting or reduced growth** with either normal or dark green colouring or yellowing.

The symptom location on the plant depends on how well the nutrient moves from older plant tissues to younger developing parts. Nutrients that can be moved readily by the plant (mobile nutrients) to younger developing tissue are nitrogen, phosphorus, potassium and magnesium. Deficiency symptoms for these nutrients are usually first expressed in the older leaves. The entire plant may develop symptoms, if the deficiency is severe. Nutrients that are not easily moved by the plant from older, developed plant parts into younger tissue are sulphur, calcium and all of the micro-nutrients. Deficiency symptoms for immobile nutrients are usually first expressed in the growing points and youngest leaves. The following is a generalized key to commonly expressed nutrient deficiency symptoms.

I. Effects occur mostly on older or lower leaves of plant; effects generalized or localized.

A. Whole plant more or less uniformly affected; may exhibit drying or firing in lower leaves :

i) Plants light green; chlorosis in lower leaves (progresses down the midrib), more or less drying or firing of lower leaves; plants may be stunted or woody; stalks short and slender, if element is deficient in later stages of growth (Nitrogen).

ii) Plants dark green, young leaves appearing abnormally dark green; stems and leaves usually highly pigmented with purplish red, especially near end of shoots; stalks short and slender, if element is deficient in later stages of growth. Fruiting often delayed. Vegetative growth less than normal (Phosphorus).

B. Effects on plants localized; mottling or chlorosis with or without spots of dead tissue on lower leaves; little or no drying up of lower leaves :

i) Mottled (often prominently) or chlorotic leaves; may redden as with cotton; sometimes with dead spots; tips and margins turned or cupped upward; stalks slender (Magnesium).

ii) Mottled or chlorotic leaves with large or small spots of dead tissue. Plants perhaps not particularly stunted; stalks may be slender; leaves, especially the older ones, scorched and dying at tip and outer margins. Leaf margins often crinkled and curled. Corn stalks may be brittle (browning of tissue evident in split joints, especially towards base of plant) with cobs not filled to the ends (Potassium).

II. Terminal buds or younger leaves affected; symptoms localized

A. **Terminal bud death common,** following are the appearance of distortions at tips or bases of younger leaves :

i) Young leaves of terminal bud at first typically hooked, finally drying back at tips and margins, so that later growth is characterized by a cut-out appearance at these points; stalk finally dies at terminal bud. Tips of unfolding leaves gelatinize, sticking together, when dry as in corn **(Calcium)**

ii) Young leaves of terminal bud becoming light green at bases, with final breakdown here, in later growth, leaves become twisted; stalk finally dies back at terminal bud. Sometimes, a distinct cupping of young leaves. With fleshy tissues, often internal browning and death **(Boron)**

B. Terminal bud commonly remains alive.

i) **Young leaves permanently wilted (wither tip effect)**, without spottingor marked chlorosis; twig or stalk just below tip and seed head often unable to stand erect in later stages, when shortage is acute **(Copper)**

ii) **Young leaves not wilted**; chlorosis is present with or without spots of dead tissue (necrosis).

- **Necrosis not commonly present.** Plants not particularly stunted.Leaves pale green, veins often somewhat lighter in colour than interveinal area. Often difficult to distinguish from nitrogen deficiency **(Sulphur)**
- **Necrosis commonly present, in spots scattered over the leaf,** Interveinal tissue yellowish, veins green **(Manganese)**
- **Necrosis may be present, often confined to leaf tip or margins,** Interveinal tissue yellowish, veins often green but may become lighter in colour later **(Iron)**
- **Necrosis may be present, generally within interveinal tissue surrounding midrib**, veins remain green, younger leaves yellowish (striping in grasses) or even white (white bud in corn) **(Zinc)**

Deficiency symptoms of plant nutrients

Nitrogen

Nitrogen being a mobile element within a plant, its deficiency results in movement of N from older to younger (upper) leaves. As a result the older leaves turn yellow in colour:

1. In young plants, growth is stunted with yellowish green leaves. Yellowing of leaves is due to the collapse of chloroplasts resulting in decrease of chlorophyll content.
2. Yellowing always starts from the older leaves and spreads to young ones.
3. Severe N deficiency affects the leaf tissue to become dry and necrotic.
4. Older leaves are shed prematurely.
5. Root growth is affected and branching is restricted.
6. Shoot become short, thin with upright growth and spindly appearance.
7. Flowering is reduced.
8. In cereals, tillering is poor, number of ears per unit area and number of grains per ear head is reduced.

Phosphorus

Generally, the symptoms of P deficiency appear in the older leaves :

1. Young plants are stunted with dark blue green leaves.
2. Stems become slender
3. Root system is limited. Primary and secondary roots elongate in length with short tertiary roots.

4. In many plant species, P deficiency induces the formation of anthocyanin pigment and the leaves acquire purple colour primarily along margins and on the lower stalk (e.g. Maize)
5. The formation of fruits and seeds is depressed and ripen slowly.
6. Plant often dwarfed at maturity.
7. In potato, tubers may develop rusty lesions in the flesh.

Potassium

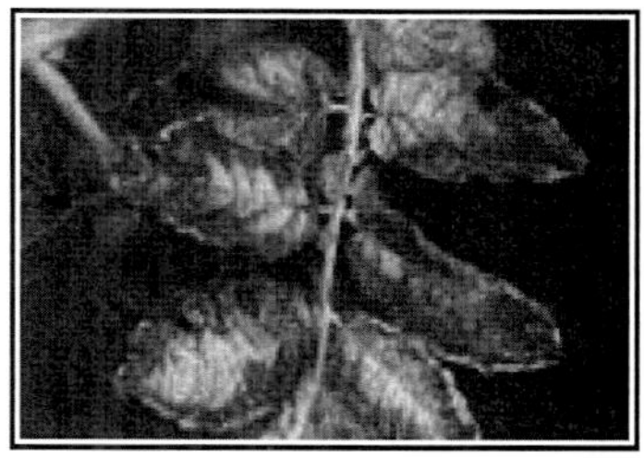

1. The most characteristic symptom of K deficiency is tip and marginal scorching of most recently matured leaves.
2. In barley which is the most susceptible of the cereals, numerous small brown areas develop in the areas between the veins.
3. Roots are slender and poorly developed in sugar beet.
4. In certain temperate fruit trees, K deficiency may result in severe **"die-back"**
5. Resistance of plants to infection by bacterial and fungal pathogens is reduced.
6. Putriscine (Diamine) accumulates under K^+ deficiency.

Calcium

1. Symptoms of Ca deficiency appear earliest and severely in meristematic regions and young leaves because it is immobile one.
2. Breakdown of meristematic tissues in stems and roots occur.
3. Roots poorly developed, lack fiber and may appear gelatinous.
4. The apical bud "dies off" and thus, small branches arise from lower leaf axils giving the plant bushy appearance.

5. Little or no fruiting.
6. Calcium deficiency causes "**bitter pit**" disease in apple and "**blossom end rot**" in tomato and water melon.

Magnesium

1. Interveinal yellowing or chlorosis occurs in older leaves and under severe deficiency the areas become necrotic.
2. In cotton, Mg deficiency induces formation of anthocyanin pigments and a reddish colouration of leaves during winter.
3. In citrus, the chlorosis commences at the tips and margins of the leaf and spread in ward towards midrib by leaving inverted V shape at the bottom of the leaf.

4. In general, fruit trees are susceptible to Mg deficiency and show varied patterns of chlorosis, necrosis and pigmentation of old leaves which shed prematurely.

Sulphur

1. Chlorotic symptoms first appear in younger most recently formed leaves.
2. Shoot growth is more affected than root growth.
3. Leaves become narrow and chlorotic in cruciferae.
4. Stems often become slender.

Iron

1. Being relatively immobile, Fe deficiency appears first on the younger leaves of the plant.

2. In most species, chlorosis is interveinal and a fine reticulate pattern can be observed in the newly formed leaves, the darker green veins contrasting markedly against a lighter green or yellow background. The youngest leaves may often be completely white and totally devoid of chlorophyll.
3. In the leaves of cereals, the deficiency is shown by alternate yellow and green strips along the length of the leaf.
4. In fruit trees, iron deficiency is associated with high leaves of $CaCO_3$ in the soil, where it is called "**lime induced iron chlorosis**".

Manganese

1. Mottled chlorosis with veins green which appears first on younger leaves.
2. Stem becomes yellowish green, often hard and woody.
3. In groundnut leaves, the veins are surrounded by a definite zone of non-chlorotic interveinal tissue.
4. Characteristic symptoms of Manganese deficiency in certain crops have been given specific names. They are **Grey speck of oats, speckled yellow of sugar beet, Marshy spot of peas, Pahala blight of sugarcane,** etc.

Boron

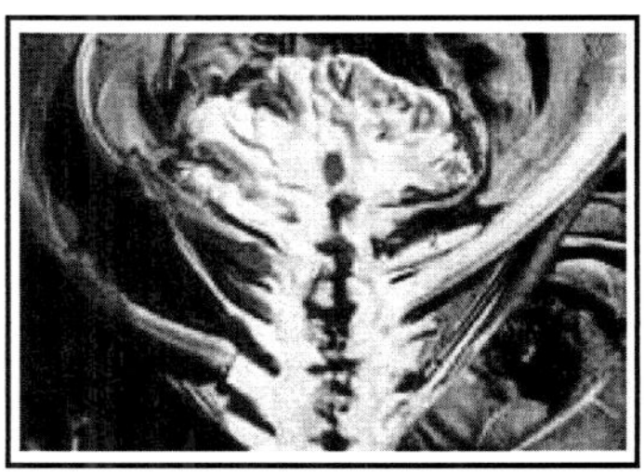

1. B is immobile in most plant species and symptoms appear first at tops and roots.
2. The youngest leaves are misshaped, wrinkled and are thicker and of a dark bluish green colour. Shed prematurely.
3. Plants dwarfed, stunted, flower development and seed production usually impaired.
4. In fruit trees, the bark of the stem may become rough and show splitting. Fruits often develop corky areas internally.

5. Roots exhibits swollen root tips.
6. Characteristic symptom of B deficiency in certain crops are :
   a) Heart rot in sugar beet
   b) Hollow stem in cauliflower
   c) Brown heart of turnip
   d) Corky pith of apples
   e) Hard fruit of citrus

Copper

1. Young leaves show chlorosis.
2. Terminal leaves, shoots are wilted and distorted, frequently followed by death resulting in development of several auxiliary buds.
3. Cereals show bushy growth with white twisted tips and reduced panicle formation.
4. Poor or no heading in cabbage and lettuce.
5. In citrus, Cu deficiency leads to **exanthema or die-back**.

Zinc

1. Little leaf and rosette are the common symptoms of Zn deficiency.
2. In maize, Zn deficiency results in white or yellow emerging leaves, a condition called **White bud.**
3. Leaves chlorotic and necrotic, young growth first affected, resulting, premature shedding.
4. In Rice, Zn deficiency causes **'Khaira disease'**
5. In monocots and particularly in maize, chlorotic bands form on either side of the midrib of the leaf.

Molybdenum

1. Symptoms of Mo deficiency are similar to those of N deficiency because in the absence of Mo, nitrate is not metabolized.
2. Leaves show marginal scorching and rolling or cupping.
3. In extreme deficiency, the leaf lamina is not formed and only the rib is present in cauliflower, which is called **Whiptail disorder.**

Chlorine

1. Chlorosis of younger leaves and an overall wilting of the plant.
2. In some plant species, like tomato, leaves show chlorotic mottling, bronzing and tissue necrosis.

Silicon

1. Withering of leaves and wilting of plants, which are called weeping willow habit of growth.
2. In cereals and grasses, necrotic spots develop on leaves.

Nutrient deficiencies and some specific names

Calcium	Bitter pit in apple, Blossom end rot in tomato
Manganese	Grey speck of oats, speckled yellow of sugar beet, Marsh spot of pea Pahala blight of sugarcane, Frenching of tungtree
Boron	Heat rot in sugar beet, Browning or hollow stem in cauliflower, Top sickness of tobacco, Brown heart of turnip, Cracked stem of celery, Corky pith of apple, hard fruit of citrus
Copper	Exanthema or dieback in citrus
Zinc	White bud in maize, Khaira disease in Rice
Molybdenum	Whiptail in cauliflower

Indicator plants

Deficiency	Crop
Nitrogen	Maize, Sorghum, Leguminous plants
Phosphorus	Tomato, Maize, Lucerne, Cereals, Duranta
Potassium	Maize, Lucerne, Cotton, Potato, Banana, Cucurbits
Sulphur	Lucerne, Clover, Cereals, Tea
Zinc	Maize, Tomato, Potato, Beans, Citrus
Copper	Citrus
Iron	Ornamental plants, Ixora, Acacia, Eucalyptus, Gooseberry, Securmanis
Boron	Lucerne, Coconut, Guava
Manganese	Citrus
Molybdenum	Cauliflower, Cabbage

Chapter 43

Sampling, Handling and Preparation of Plant Samples

Plant analysis is the determination of concentration of an element in a sample from a particular part of a plant sampled at a particular stage of morphological development. Plant analysis is based on the principle that the concentration of a nutrient within the plant is an integral value, if all the factors that have integrated to affect it.

There are two general classes of plan analysis:

1. Total chemical or quantitative analysis
2. Semi-quantitative or rapid tissue analysis

Each may be used at different stages of plant growth and on different parts of the plant or on the entire plant.

The total analysis measures the elements that are incorporated as complex compounds and those that are present as soluble ions in the plant sap. The determination is made by :

1. **Gravimetric method:** Separate the element chemically by precipitation and weigh it.
2. **Flame photometric method:** By measuring the amount of light emitted by the elements after excitation.
3. **Atomic Absorption Spectrophotometry method:** By measuring absorption of characteristic light by elements in their ground state.
4. **Absorption Spectrophotometry or Colorimetry:** The amount of monochromatic light absorbed by the coloured compound.

In rapid tissue tests, one measures the unassimilated soluble constituents of the plant cell sap, which are yet to be assimilated.

The value of plant analysis and proper interpretation depends mainly on collection of proper sample, its careful handling during preparation and analysis. Errors committed during these stages result in erroneous interpretations and to the grower and sometimes, even disastrous. Hence, it is very important that persons employing plant analysis be familiar with recommended sampling, handling and analysis procedures.

Sampling

This is the most crucial step in analysis. The usefulness and validity of accurate chemical analysis depends on intelligent and realistic approach to the problem of obtaining a reliable plant sample. If the sample obtained is not a true representative, all the careful and costly work put into subsequent analysis will be of no value and results will not be valid.

Which plant part to be sampled?

The nutrient composition of the plant species is not a fixed entity. It varies from time to time in plants, from soil to soil and even from species to species (Jones, 1970). Hence, a specific plant part should be chosen at definite stage or plant growth. The physiologically matured plant part should be selected, as it will not undergo rapid changes in nutrient composition. The plant part selected and the time of sampling should be such that a good correlation exists between the concentration of the nutrient and yield (Bates, 1971). The plant part to be selected and the time of sampling varies from species to species.

How much to sample?

The number of plants to be sampled varies from species to species and it depends on a number of conditions, such as general condition of the plants, soil homogeneity, and purpose for which samples are to be collected. For most of the field crops, collecting 2 leaves per plant from 20-40 plants at random is recommended.

When to sample?

Sampling should bedone just prior to or at the time the plants begin reproductive stage of growth. Sampling is not recommended, when plants are diseased, damaged by insects or mechanical (agricultural implements). Plants under nutrient stress, moisture stress and cold stress should not be collected. Dead plant tissue and plant parts covered by soil or dust should not be collected (Reisenauer, 1956).

Preparation of samples

Different steps to be followed are :

a) After sampling, the samples should be transferred to muslin bag and transported to the laboratory as quickly as possible. For long distance transportation (>2 days), the samples should be transported under refrigerated conditions.

b) Wash the samples by sponging with a piece of cotton wool using 0. 1% detergent solution to remove waxy coating, soil and dust particles on leaves. Then wash with N/10 (0. 1 N) HCl to remove metallic contaminants, which are added through spraying of pesticides. Washing should be done as quickly as possible. Delayed contact of leaf with detergents and acids may leach out nutrients, such as K andCa from the samples. Wash the samples with running water and rinse with deionized/distilled water, wipe out the extra moisture and place them in paper bags.

c) Dry the samples in an oven at 65-70°C for 48 hrs. to remove the moisture and to inactivate the enzymes.

d) Grind the samples at this temperature using a Wiley mill (stainless steel wheels) or a homogeniser with stainless steel blades (mixers) to prevent contamination with metal ions, such as iron, zinc and copper. Grinding mills made of hard plastics are preferred. After grinding, mix the samples thoroughly and dry again at 70 °C to remove the moisture.

e) Store the samples in screw type plastic or glass bottle to prevent contamination.

Microwave digestion for nutrient estimation

Principle

Microwaves are electromagnetic radiation of frequencies ranging from 100 to 1,00,000MH_Z generally, the cooking ovens have a range of 2450MH_z (megahertz). In the oven cavity, the microwaves are reflected from the metal walls to the solution container, where the solution absorbs the energy in the radio waves. This energy is converted to heat by the molecules trying to align themselves with the fast moving radio waves.

Actually, with a frequency of 2450 MH_Z, the solution molecules are changing directing four billion nine million times a second and a great deal of molecular friction is developed. Friction generates the heat to boil the solution. The radiation passes through glass or plastic and doesn't couple with the container, as is the case with conventional heating.

Generally, metals reflect the wave energy. Glass, paper and most plastic allow the wavesto pass through them without absorbing energy. Hence, metal containers are not used in microwave oven. The organic matter (made of H_2O molecules) also have high energy absorption. A typical microwave oven uses a magnetron to produce power of the order of a Kilowatt. The radiation is beamed by wave guides into a cavity designed to resonate at that frequency, so producing an even energy distribution.

Reagents

1. Conc. Nitric acid (specific gravity of 1. 42 g cc^{-1})
2. Hydrogen peroxide (30%)
3. Conc. Hydrochloric acid.

Procedure: (Digestion in microwave digester)

1. Transfer 0. 5 g of sample to Teflon digestion vessel (more, if concentration of nutrients is less). Add 0. 5 ml of H_2O_2 and 2-4 ml of HNO_3acid and swirl the contents.
2. Screw on the caps, load digestion vessel on turn table and place it in the oven properly. Switch on the turn table and check to ensure its smooth rotation.
3. Enter in exhaust and digest time and power in the programme and the programme is initiated only after ensuring the power supply to the exhaustion unit (programme depends on the type of sample).
4. At the end of digestion, cool the contents in a plastic tub, remove the caps in a fume hood and rinse the cap and sides of digestion vessel with water: filter, if necessary and make up the volume to 100 ml.

Chapter 44

Estimation of Total Nitrogen in Plant Samples (Micro Kjeldahl's Method)

Principle

A known quality of powdered sample is digested with conc. H_2SO_4 in the presence of digestion mixture by heating in fume chamber. Under these conditions, $[NH_4]_2SO_4$ is formed from protein, nitrogen and non-protein nitrogenous compounds. The mixture is distilled and NH_3 released is trapped in boric acid and which is then back titrated with Std. acid.

1ml of 1 N H_2SO_4 = 0.0141g of N

Using the above relationship, the concentration of N in the plant sample is calculated.

$$\text{Protein + Non-Protein compounds} \longrightarrow (NH_4)_2SO_4$$

$$(NH_4)_2SO_4 + NaOH \longrightarrow Na_2SO_4 + NH_3 + H_2O$$

$$NH_3 + H_3BO_4 \longrightarrow (NH_4)^z - H_2BO_3$$

$$(NH_4)^+ - H_2BO_3 + H_2SO_4 \longrightarrow (NH_4)_2SO_4 + H_3BO_3$$

Reagents

1. **Concentrated H_2SO_4**
2. **40% NaOH solution:** Dissolve 40 g of NaOH in about 80 ml water and make up the volume to 100 ml.
3. **Boric acid (4%):** Dissolve 40 g of boric Acid in hot distilled water, cool and then make up the volume to 1 litre.
4. **Mixed Indicator:** Mix 0.1g of bromocresol green with 0.07 g of methyl red and dissolve the mixture in 100 ml ethanol.

5. **Standard H_2SO_4 (0.05 N):** Dilute 1.4 ml of conc. H_2SO_4 to 1000 ml and then standardize it against std. Na_2CO_3 solution.
6. **Digestion mixture:** Mix 100 parts of K_2SO_4, 20 parts of $CuSO_4$ and 1 part of Selenium (Se) powder and grind it to get fine powder.
7. **Standard Na_2CO_3 (0.1 N):** Dissolve 5.3 g of the dried Na_2CO_3 in some volume of distilled water (500 ml) and then make up the volume to 1 litre.

Materials

Kjeldahl's digestion flask, volumetric flask, micro-distillation apparatus, pipette and burette.

Procedure

1. Weigh 1g of powdered sample on to the sheet of paper, prepare a packet and transfer to Kjeldahl's digestion flask. Add 2 to 3 g of digestion mixture and about 10-15 ml of conc. H_2SO_4. Digest the contents in fume cupboard (digestion chamber) over a low flame till a light bluish green residue is obtained. Then cool the contents and make up the volume to 100 ml with distilled water.
2. Pipette out 10ml of the digested sample into a micro distillation flask and add about 10ml distilled water.
3. Take about 20-25 ml of 4 per cent boric acid in a receiving flask and connect to receiving tube in such a way that the delivery (in receiving flask) is dipped in the boric acid.
4. Add about 10 ml of 40 per cent NaOH to the distillation flask and distil the contents.
5. After complete distillation, disconnect the receiving flask and titrate the concentrate of receiving flask against Std. acid till it becomes pink in colour.
6. From the burette reading calculate the per cent N.

Calculations

$$\text{Per cent N} = \frac{\text{T. V x N of acid x 0.014 x Volume of digested sample}}{\text{Wt. of sample x Aliquot taken}} \times 100$$

Note

1. To trap the NH_3, we can use the Std. H_2SO_4 instead of boric acid. The volume and the concentration of the acid taken for trapping of NH_3 is very crucial in quantification. The whole content after trapping should be titrated against Std. using methyl red indicator.

2. Per cent crude protein = Per cent N x 6.25

   (This does not holds good for wheat and its products).

Chapter 45

Estimation of Phosphorus in Plant Samples (Vanadomolybdo Phosphoric Yellow Colour Method in HNO_3 Acid System)

Principle

In the presence of Vu z and Mov z, orthophosphates form a yellow coloured phospho-vanadomolybdate complex, which shows an optimal absorption at wavelength of 430 nm. Once the absorption is measured, the absorption measurement (optical density) can be related directly to the concentration of P by comparing with the measured absorptions of a series of standards.

Reagents

1. **Phosphate standard (50 ppm of P):** Dry the Potassium dihydrogen phosphate (KH_2PO_4) at 40°C and dissolve 0.2195 g in about 400 ml of distilled water present in a 1 litre volumetric flask. Then add 25 ml of 7 N H_2SO_4 and make up the volume to 1 litre with distilled water.
2. $HN0_3$ – Vandatemolybdate reagent:
   a) **Solution A:** Dissolve 25g of ammonium molybdate in 400ml of distilled water.
   b) **Solution B:** Dissolve 1.25g of ammonium meta-vandate in 300ml of boiling distilled water. Cool it and add 250ml of conc. $HNO_{3.}$

Finally mix the solution A and B and dilute the mixture to 1 litre with distilled water.

Materials

Spectrophotometer, volumetric flasks, pipettes, *etc.*

Procedure

Preparation of standard curve

a) Take 0, 0.5, 1.0, 2.5, 5.0, 7.5 and 10.0ml of 50 ppm of solution in separate 50 ml volumetric flask.

b) Add 10 ml of HNO_3-Vanadomolybdate reagent to each flask.

c) Make up the volume and mix it.

d) After 30 minutes, record the intensity of the colour (OD) at 430 nm

e) Plot the OD against concentration and draw the curve.

Sample

a) Pipette out 5 ml of digested sample into a 50 ml volumetric flask and add 10 ml of HNO_3- vanadomolybdate reagent.

b) Make up volume to 50 ml and mix.

c) After 30 minutes, record the intensity of the colour at 430 nm.

d) Compare the unknown sample OD with standard curve and calculate % P of the sample.

$$\text{Per cent P} = \frac{\text{Graph ppm x Volume of digested sample}}{10^6 \text{ x Wt. of sample}} \times 100$$

If dilution is done,

$$\text{Per cent K} = \frac{\text{Graph ppm x Volume of digested sample x Volume made up}}{10^6 \text{ x Wt. of sample x Aliquot taken}}$$

Chapter 46

Estimation of Potassium in Plant Samples (Flame Photometry)

Principle

In flame photometry, also known as flame emission or flame atomic emission. The sample in solution, is sprayed into a flame to vaporize, atomize, and excite the sample. The excited atoms of the element of interest emit light at certain discrete wavelength, which are characteristic of that element. Light of wavelength of interest is separated from the reminder of emitted radiations and its intensity is measured. The intensity measurement can be related directly to the concentration of the element of interest usually by comparing with the measured intensities of a standard or series of standards.

Reagents

Potassium standard 100ppm of K: Dissolve 0.1910 g of KCl in some volume of distilled water and then make up the volume to 1 litre.

Materials

Flame photometer, pipette, volumetric flask, etc.

Procedure

a) **Preparation of standard curve:** Take 0, 1, 2, 3, 4, and 5 ml of 100 ppm K solution in a separate 50 ml volumetric flasks. Make up the volume to 50 ml with distilled water and mix well. After that adjusting the needle of flame photometer to zero by feeding blank, adjust the needle to 100 by feeding maximum concentrated K solution. Then feed standards to record the flame photometer readings. Plot flame photometer readings versus concentrations of standards and draw the standard curve.

b) Feed the digested sample solution to the flame photometer and record the readings (if dilution is required, one should do the dilution before feeding to the instrument).

Compare the unknown sample readings with standard curve to determine the % K in the plant sample.

$$\text{Per cent K} = \frac{\text{Graph ppm x Volume of digested sample}}{10^6 \text{ x Wt. of sample}} \times 100$$

If dilution is done,

$$\text{Per cent K} = \frac{\text{Graph ppm x Volume of digested sample x Volume made up}}{10^6 \text{ x Wt. of sample x Aliquot taken}} \times 100$$

Chapter 47

Estimation of Calcium and Magnesium in Plant Samples (EDTA - Titration Method or Versenate - Titration Method)

Principle

Ethylene diamine tetra acetic acid will be assigned the formula H_4Y: The sodium salt is, therefore, Na H_2Y and affords the complex-forming ion $HY^{=}$ in aqueous solution. The usefulness of EDTA has a titrant is due to the presence of four or six atoms, which are available for coordination to a metal cation is such a way that five membered rings are produced; 1:1 complexes are formed and these are the most important ones. The reaction with cation e.g. M^{+2}, may be written as:

$$M^{++} + H_2 \longrightarrow MY^{=} + 2\,H^{+}$$

$$M^{4} + H_2Y \longrightarrow MY + 2H^{+}$$

$$Mn^{+} + H_2Y \longrightarrow (MY)^{(n-4)} + 2H^{+}$$

For other cations, the reaction may be expressed as:

$$M^{+3} + H_2Y \longrightarrow MY^{-} + 2H^{+}$$

One gram ion of the complex forming $H_2Y^{=}$ reacts in all cases with one gram ion of the metal and each case, two hydrogen ions are formed. The resulting complexes have similar structure but differ from one another in the charge that they carry.

Reagents

1. **0.01 N Standard EDTA solution**: Dissolve 2 g of EDTA in 900 ml of distilled water and make the volume to 1 litre.
2. **Standard Ca solution:** Dissolve 0.6005 g portion of pure dried $CaCo_3$ in minimum of 0.2 N HCl (5.0 ml Conc.), solution is boiled to expel the CO_2 and dilute to 1 litre.

3. **10% NaOH solution:** Dissolve 10 g portion of NaOH in about 90 ml distilled water and dilute to 100 ml.
4. **Murexide indicator:** Mix 0.2 g portion of murexide with 40 g of powdered K_2SO_4 and powder to get homogenized fine powder.
5. **NH_4OH + NH_4Cl buffer (pH 10):** Add 142 ml of NH_4OH to 17.5 g of NH_4Cl and dilute to 250 ml with distilled water.
6. **Erichrome black T indicator:** Dissolve 0.2 g of EBT in 15 ml of triethanolamine and 5 ml of absolute ethanol.

Materials

Porcelain basin, glass rod, burette, pipette, etc.

Procedure

A. Determination of calcium alone

Take 5 ml of digested sample in a porcelain basin and dilute it with distilled water (about 25 ml). Add sufficient quantity of 10% NaOH (pH of sample be more than 12). Add about 0.5 g of murexide indicator. Titrate the contents against standard EDTA with stirring until it become violet in colour. Note down the burette reading.

B. Determination of Ca + Mg

Take another 5 ml of digested sample in a porcelain basin and dilute with 25 ml of distilled water. Add sufficient quantity of buffer solution (about 10 ml). Add 3-5 drops of EBT – indicator and then titrate the contents against standard EDTA by stirring till it turns blue. Note down the burette reading. To get burette reading of Mg substract burette reading of Ca with burette reading of Ca + Mg.

$$\text{Per cent Ca} = \frac{\text{B.R. of EDTA xN of EDTA x 0.02 x Volume of digested sample}}{\text{Aliquot taken x Weight of sample}} \times 100$$

$$\text{Per cent Mg} = \frac{\text{B.R. of EDTA x N of EDTA x 0.012 x Vol. of digested sample}}{\text{Aliquot taken x Weight of sample}} \times 100$$

Note

If the concentrations of interfering ions are more in the sample, then 1 or 2 drops of 1% KCN or NaCN should be added to the sample before titration.

Chapter 48

Estimation of Sulphur in Plant Samples (Turbidometry)

Principle

'S' is released into solution from plant material by digesting with diacid mixture (HNO_3+$HClO_4$). The released 'S' in the solution is precipitated by Ba^{+2} ions as $BaSO_4$. This gives turbidity to the solution, which is then determined by turbidimeter.

Turbidimetric analysis is based on measuring the weakening intensity of a luminous flux, when it passes through a solution containing particles in suspension. The intensity decreases owing to absorption and scattering of light.

$$\text{Turbidence} = \log\frac{Io}{It} = K \times l \times C$$

Where:

I_0 = intensity of incident ray

I_t = intensity of transmitted ray

l = length of absorbing medium

C= concentration of particles

K = constant

Using the above relationship, the concentration of 'S' of plant is calculated by comparing the unknown sample turbidance with standard curve.

Reagents

1. **Acid seed solution, 20 ppm 'S' in 6 N HCl:** Dissolve 0.109 g of K_2SO_4 in some volume of 6 N HCl and make up the volume to 1000 ml with 6 N HCl.

2. **$BaCl_2$ crystals** (20-60 mesh size).
3. **Stabilizing reagents:** 0.25% Gum acacia solution or mix 20 ml of glycerol with 80 ml of alcohol.
4. **Standard 'S' solution, 100 ppm S:** Dissolve 0.544 g ofK_2SO_4 in 1 litre of solution.

Materials

Spectrophotometer or Turbidometer, pipette, volumetric flasks, etc.

Procedure

a) For Standard Curve

1. Take 0, 2.5, 5, 10, 15, and 20 ml of 100 ppm 'S' solution in separate 50 ml volumetric flasks.
2. Add to each 1 ml acid seed solution and 1 ml stabilizing agent.
3. Dilute the contents to about 40 ml.
4. Add to each, 1g of $BaCl_2$ crystals and make up the volume to 50 ml.
5. Mix well and record the turbidity of these solutions at 420 nm within 20 min.
6. Plot the turbidance against the concentration of these solutions and draw the curve.

b) Plant sample

1. Take 10 ml of digested sample in a 50 ml volumetric flask.
2. Add one ml of acid seed solution and 1ml of stabilizing reagent.
3. Dilute the contents to about 40 ml and then add 1 g $BaCl_2$ crystals.
4. Make up the volume and mix it well.
5. Record the turbidity or % transmission at 420 nm within 20 min.
6. Compare the turbidity with standard curve in order to calculate % S in the sample.

$$\text{Per cent S} = \frac{\text{Graph ppm x Vol. of digested sample volume made up}}{10^6 \text{ x weight of sample x aliquot taken}} \times 100$$

Note: If the concentration of 'S' is more in plant sample, then there is no need of using acid seed solution.

Chapter 49

Estimation of Micro-nutrients in Plant Samples

I. Determination of Fe, Mn, Zn and Cu

Make a suitable dilutions of tri/ di acid extract and feed the sample to AAS having appropriate hallow cathode lamps after getting values for standard solutions.

II. Boron

Pipette out suitable aliquot of tri/ di acid extract and develop colour using curcumine/carmine method and measure O.D using a spectrophotometer.

Chapter 50

Estimation of Chlorophyll Content of Leaves

Chlorophyll is the green pigment found in plants and it's vital for photosynthesis.

Why to measure the chlorophyll content in the field?

Using a portable chlorophyll meter in the field allows near real-time data to be collected. It is a relatively simple method that, although indirect, allows measurements to be made *in situ* and in a time series. A chlorophyll meter is an ideal choice for researchers, who are interested in detecting the presence or absence of chlorophyll, and measuring relative fluorescent changes that can be used as an indication of increasing or decreasing concentrations.

Using a chlorophyll meter allows you to identify temporal and vertical patterns of leaf biomass in plants. This method is an effective measure of trophic status, a potential indicator of maximum photosynthetic rates of plants. Another benefit of measuring chlorophyll is quickly to pulsed nutrient inputs that might otherwise go undetected by regular nutrient sampling.

The chlorophyll present in the plant leaves is closely related to the nutritional condition of the plant. The chlorophyll content will increase in proportion to the amount of nitrogen present in leaf.

Instruments used to measure the chlorophyll content in plants are :

1. Chlorophyll meter
2. Leaf colour chart (LCC)
3. Green seeker

I. Chlorophyll meter

Principle

The meter works by emitting two frequencies of light, one at a wavelength of 660 nm (red) and one at 940 nm (infrared). Leaf chlorophyll absorbs red light but not infrared, the difference in absorption is measured by the meter and termed "Optical Density Difference," ODD. Therefore, the unit of measurement is ODD, a ratio that is provided by the meter. The value does not give an actual chlorophyll or nitrate count rather than saying, "There is this number of chlorophyll," the meter value can be interpreted as, "This is a ratio of reflection v/s absorption".

Additionally, plants cease to create chlorophyll once a certain threshold has been reached. Consequently, the meter cannot indicate an abundance of nitrate, only a possible deficiency. In terms of estimating nitrate content, the measured value is most useful, when compared to a well-fertilized control group.

Nitrogen management with chlorophyll meter

Nitrogen (N) management is very important issue for plant growers and for the environment. For grower, knowing the N requirement of plant enables the proper amount of N fertilizer supply to be managed. Some experiments shown that the chlorophyll meter (SPAD Meter) contributed to reductions in the use of N fertilizer by tens of per cents with no loss in yield. By optimizing the N fertilization efficiency in the field, proper N fertilizer management reduces the possibility of excessive supply of fertilizer, which can cause diseases in plants and environmental contamination. There is increasing awareness of the need for applying the proper amount of N fertilizer in respect to water contamination in both flowing streams and underground water due to nutrient leaching through the field's soil. The chlorophyll meter is playing a significant role in day-to-day development of N fertilizer application techniques.

Advantages of chlorophyll meter

- High accuracy
- Compact and lightweight for portability
- Quick and easy measurements
- Water resistant
- Low power consumption
- Small measuring area

II. Leaf Colour Chart (LCC)

The LCC is usually a plastic, ruler-shaped strip containing four or more panels that range in colour from yellowish green to dark green. LCC was developed by IRRI (International Rice Research Institute) and also by PRRI (Philippines Rice Research Institute). In India, these charts are provided by Nitrogen parameters which provide guidance to the farmers regarding improved Nitrogen (N) management. Nitrogen is applied by the farmers in rice plants at different growth stages, but the amount of N to be applied and the time of applications vary substantially. LCC helps to synchronize N application with the real time and quantity demand of the rice crop. Helps to prevent under or over usage of N in rice plants.

Critical value to apply N fertilizer

In six panel LCC, critical readings are 4 for transplanted/hybrid rice, 3.5 semi-dwarf *indica* rice and 3 for basmati rice.

How to use LCC?

1. At 14 days after transplanting (DAT) or 21 days after direct wet seeding (DAS), randomly select 10 healthy plants in your field, where plant distribution is uniform.

2. Compare the topmost, fully expanded, and healthy leaf of each of the 10 plants with the LCC. Place the middle part of the leaf on top of the LCC's colour strips for comparison. Do not detach the leaf. Take readings at same time of the day (8-10 AM). Do not expose the LCC to direct sunlight during readings. The same person should take the first upto the last LCC reading.

3. If six (6) or more of the 10 leaves have readings below the critical LCC value, apply N as given below :

   - For wet season (Kharif) non-basmati rice, use LCC critical value 4, and apply 28 kg N/ha or 1.25 bag urea per hectare.

- For wet season (Kharif) basmati rice, use LCC critical value 3, and apply 23 kg N/ha or 1 bag urea per hectare.
- For direct-seeded rice, apply 23 kg N/ha as basal, then use LCC critical value 3, and apply 23 kg N/ha or 1 bag urea per hectare.
- For Boro rice, apply 23 kg N/ha as basal, then use LCC critical value 4 and apply 35 kg N/ha or 1.5 bags of urea per hectare.

4. Repeat LCC readings every 7 days for 110-130-days rice crops and every 10 days for more than 130 days crops until first heading. Different sets of 10 leaves can be used for each weekly or 10 days readings.

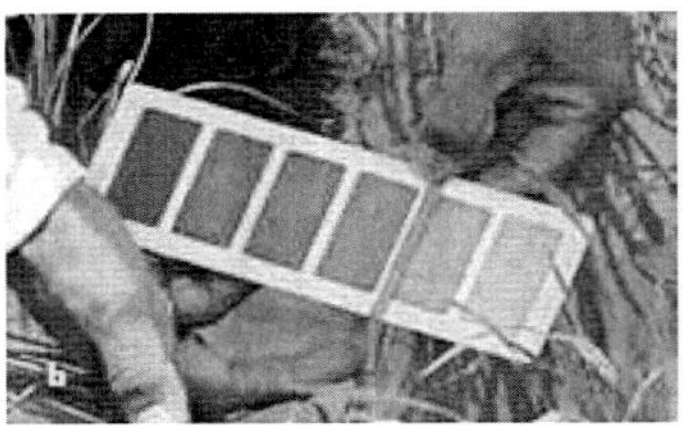

Advantages

1. Low cost
2. High production
3. Saves urea
4. Less production of greenhouse gases
5. Less infestation by pest
6. Cheap and easy to use
7. Low volatilization prevents excessive loss of nitrogen

Disadvantages

1. Less popular
2. Less availability
3. Reading may be affected by variety, solar radiation, and other nutrient deficiency.

III. Green seeker

The Green Seeker handheld crop sensor is an affordable, easy-to-use measurement device that can be used to assess the health or vigour of a crop. Readings taken by the Green Seeker handheld can be used to make non-subjective decisions regarding the amount of fertilizer to be applied to a crop, resulting in a more efficient use of fertilizer - a benefit to both a farmer's bottom line and the environment.

How the Green Seeker works: Handheld Crop Sensor works :

- Upon pulling the trigger, the sensor turns on, emits brief bursts of red and infrared light, and then measures the amount of each type of light that is reflected back at the sensor.

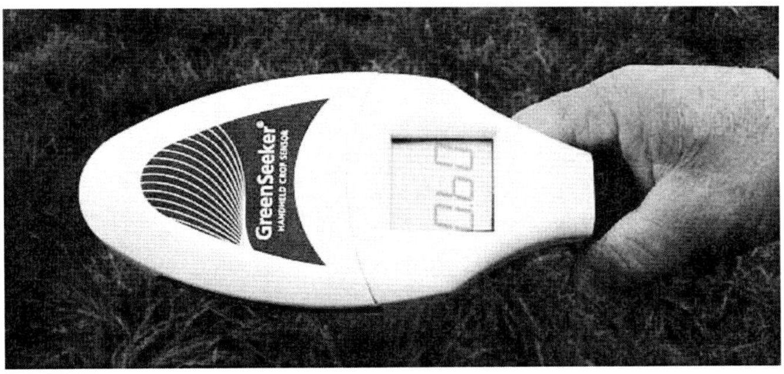

- The sensor continues to sample the scanned area as long as the trigger remains engaged.
- The strength of the detected light is a direct indicator of the plant's vigour.
- The sensor displays the measured value in terms of an NDVI reading on its LCD display screen. NDVI readings can range from 0.00 to 0.99; the higher the reading, the healthier the plant.
- To ensure the accuracy of your readings, hold the sensor 60-120 cm above the crop, when the trigger is pulled.

Advantages

- Reduce in-crop fertilizer and input costs
- Allows inputs to be adjusted during the growing season to meet plant requirements
- Only apply inputs to areas that need it
- Collect data during existing farming operations
- Fast and precise optical sensing

Chapter 51

Preparation of Acid and Alkali Standard Solutions

I. Preparation of a standard solutions of an acid

Acids are chemicals which contain one or more replaceable hydrogen atoms. Bronsted defined acids are chemicals which donate protons in a reaction. Some of the properties of acids are:

1. They are sour to taste.
2. Impart characteristic colours to certain organic dyes called indicators – they turn blue litmus red, methyl orange to pink, phenolphthalein remains colourless.
3. React with bases to form salts and water.
4. Liberate carbon dioxide from carbonates.
5. Their solutions in water are good conductors of heat and electricity.

In most of the chemical analysis, standard acids are used. But as the acids are not chemically pure and as many of them absorb moisture while weighing, we cannot prepare standard solutions of acids. Hence, they should be standardized using a primary standard base, such as sodium carbonate or borax or mercuric oxide using a suitable indicator.

Principle

A known volume of standard sodium carbonate solution is titrated against hydrochloric acid solution using methyl orange as indicator.

Reagents required

1. **0.1 N sodium carbonate solution:**Dissolve 5.3g of Na_2CO_3 (oven dried) in 1 litre of water.
2. **Approximately 0.1 N hydrochloric acid solution:** Dissolve 10ml of concentrated HCl in 1 litre of water.
3. **Methyl orange:** The indicator is obtainable either as free acid or as its sodium salt.Dissolve 0.5g of the free acid in one litre of water and filter in cold solution, if necessary,or dissolve 0.5g of the sodium salt in one litre of water and add 15.2 ml of 0.1 N HCl.Filter if necessary, when cold.

Procedure

1. Pipette out 10 ml of standard sodium carbonate solution into a 250 ml conical flask. Add50 ml distilled water and 3 to 4 drops of methyl orange indicator.
2. Titrate against hydrochloric acid solution till the colour changes to orange or faint pink.Note down the titral value. Repeat the titration two more times or till concordant values are obtained.

Calculations

ml of Std. Na_2CO_3 solution taken = 10 ml (V_2)

Normality of Na_2CO_3 solution = 1 N (N_2)

Titral value (T. V.) = X ml (V_1)

Normality of HCl (N_1) = ?

$$N_1 V_1 = N_2 V_2$$

$$N_1 \times X = 10 \times 1$$

$$N_1 = \frac{10 \times 1}{X}$$

II Preparation of a standard solution of alkali

Bases are substances which react with an acid and produces salt and water. They are also defined as substances, which accept protons in a reaction. Generally, oxides and hydroxides of metals are bases. The bases which are readily soluble in water are called alkalis and they contain OH group.

Alkalis exhibit the following properties:

1. All alkalis are caustic and soapy to touch.
2. Impart characteristic colours to indicators - turn red litmus blue, methyl orange yellow and phenolphthalein pink.
3. React with acids to give salt and water.
4. Their solutions are good conductors of electricity.

As in case of acids, bases and alkalis also are not chemically pure and some of them are hygroscopic. Hence, they should be standardized using primary standard acid chemical, such as potassium hydrogen phthalate or potassium hydrogen iodate or benzoic acid or sulphanic acid or constant boiling HCl using a suitable indicator.

Principle

A known volume of standard potassium hydrogen phthalate solution is titrated against sodium hydroxide solution using phenolphthalein indicator.

Reagents required

1. **0.1 N potassium hydrogen phthalate solution**: Dry 25g of pure potassium hydrogen phthalate salt, in an open at 105°C for half an hour and weigh exactly 20.41g, dissolve and make up to one litre.
2. **Approximately 0.1 N NaOH solution:** Dissolve 4.5g of NaOH and make up to one litre.
3. **Phenolphthalein indicator (0.5%):** Dissolve 0.5g of Phenolphthalein in 100 ml of 95% alcohol.

Procedure

1. Pipette out exactly 10 ml of potassium hydrogen phthalate solution into a 250 ml conical flask. Add 50 ml distilled water and 2 to 3 drops of phenolphthalein indicator.
2. Titrate the contents against NaOH solution till a permanent light pink colour appears. Note down the titral value and repeat till concordant values are obtained.

Calculations

ml of standard potassiumhydrogen phthalate taken = 1 0 ml (V_2)

Normality of potassium hydrogen phthalate = 1 N (N_2)

Titral value (T. V.) = X ml (V_1)

Normality of NaOH (N_1) = ?

$N_1 V_1 = N_2 V_2$

$N_1 x X = 10 \times 1$

$$N_1 = \frac{10 \times 1}{X}$$

Chapter 52

Fertilizers and its Composition

Nitrogenous fertilizers

Material	Analysis	Physical form	Handling Precautions
Anhydrous ammonia	82-0-0	Gas	• Extremely irritating to eyes and respiratory system in concentrations upto 0.07% by volume; 0.17% causes convulsive coughing; 0.5 to 1.0% is rapidly fatal after short exposure • Will burn, if a source of ignition is present • Flammable and explosive mixtures are formed with air in concentrations of 16 to 25% NH3 by volume. • Need specialized equipment for application, used primarily for agronomic crops on relatively level land.
Urea	46-0-0	Solid (prills, granules, crystalline)	• Good storage and handling properties
Ammonium nitrate	34-0-0	Solid (prills or granules)	• Strong oxidizer; readily absorbs water • Can be explosive when mixed with carbonaceous materials
UAN solutions	(28-32)-0-0	Liquid	• Normally 33 to 35% is urea and 45 to 47% is ammonium nitrate • Weighs about 11.1 pounds per gallon; • Salting out may occur at 32 F.
Ammonium sulphate	21-0-0 + 24 S	Solid (crystalline or granules)	• Used primarily as an acidifying fertilizer or source of sulphur.

Phosphatic fertilizers

Material	Analysis	Physical form	Citrate soluble	Water soluble	Handling Precautions
Diammonium phosphate	18-46-0	Solid, granule	-	100	Avoid excess dust
Mono ammonium phosphate	11-48-0	Solid, granule	-	100	Avoid excess dust
Triple superphosphate	0-46-0	Solid, granule	13	84	Avoid excess dust
Single superphosphate(SSP)	0-20-0	Solid, granule	18	78	Avoid excess dust
Ammonium polyphosphate	11-37-0 10-34-0	Liquid	-	-	Eye irritant, dike to contain spilled material

Potassium fertilizers:

Materials	Analysis	Physical form	Handling Precautions
Potassium chloride	0-0-60	Solid, granule	Avoid excess dust
Potassium sulphate	0-0-50 + 18	Solid, granule	Avoid excess dust
Potassium nitrate	13-0-44	Solid, granule, crystal	Avoid excess dust

Complete mixed fertilizers:

Mixed Materials	Per cent			Pounds of fertilizer to supply one pound of primary nutrient		
	N	P_2O_5	K_2O	N	P_2O_5	K_2O
19-19-19	19	19	19	5.3	5.3	5.3
15-15-15	15	15	15	6.7	6.7	6.7
10-10-10	10	10	10	10	10	10
6-12-12	6	12	12	16.7	8.3	8.3
12-24-24	12	24	24	8.3	4.2	4.2
5-10-15	5	10	15	20	10	6.7
18-46-0	18	46	0	5.5	2.2	NA
11-37-0*	11	37	0	9.1	2.7	NA

*Liquid material, weighs about 12 lbs. per gallon or 167.2 gallons per ton.

Secondary and micro-nutrient fertilizers

Nutrient materials	Nutrient contents (%)	Pounds of materials for one pound nutrient
Magnesium		
Magnesium Oxy-Sulphate (granular)	36	2.8
Dolomitic Limestone*	9	11.1
Magnesium Sulphate (Epsom Salts)	9	11.1
Magnesium-Potassium Sulphate	11	9.1
Sulphur		
K-Mag (Sul-po-mag)	21-22	4.5-4.7
Elemental Sulphur	85-100	1.0-1.2
Calcium Sulphate (Gypsum)	15-18	5.5-6.7
Ammonium Sulphate	24	4.2
Boron		
Borax	10-15	6.7-10
Calcium		
Calcitic Limestone*	35	2.8
Calcium Sulphate (Gypsum)	22.5	4.4
Bone Meal	15	6.7
Iron		
Iron Sulphate	40	2.5
Iron Chelates (soluble powder)	20	5.0
Manganese		
Manganese Oxy-sulphate	28	3.6
Manganese Chelates (soluble powder)	20	5.0
Zinc		
Zinc Oxy-Sulphate	36	2.8
Zinc Chelates (soluble powder)	25	4.0

Chapter 53

Important Terms and Definitions

Primary standard: A chemical of such purity that can be weighed out and used to standardize a solution to a very high degree of accuracy

Mass: The mass of a body is the quantity of matter it contains.

Weight: The weight of a body is the gravitational force exerted on it.

Litre: A litre is the volume occupied by 1 kilogram of water at the temperature of its maximum density (at 4^0C i.e., liter = 1000.027 cc)

Important oxidizing agents used as Std. solutions: Potassium dichromate, Potassium permanganate, Potassium ferricyanide, Iodine, Ferric sulphate, Potassium iodide and Potassium bromate.

Important reducing agents used as Std. solutions: Ferrous sulphate, Oxalic acid, Sodium oxalate, Sodium thiosulphate, Titanous chloride and Sodium arsenite.

Acid: An acid is an ion or molecule which can donate one or more protons. A strong acid losses protons more easily than a weak acid.

Base: A base is an ion or molecule which can accept protons. A strong base gains proton more easily than a weak acid.

Law of Mass action (Guildberg and Waage): The rate of chemical reaction is directly proportional to the product of the formula weight, concentration of the reacting substances, each raised to the power indicated by the number of ions or molecules appearing in the balanced equation for reaction.

pH-Buffer Solution: A buffer solution is one which resists the change in hydronium ion concentration, when an appreciable amount of either a strong acid or base is added to the solution.

Electrolytes: When dissolved is water or any other ionizing medium, produce electrically charged particles called ions.

Osmosis: The transfer of solvent molecules through a semi-permeable membrane to cause the dilution of a solution is called osmosis. The osmotic pressure of a solution is the external pressure, which must be applied to the solution, when it separated from pure solvent by a semi-permeable membrane, to prevent the passage of the solvent molecules through the membrane.

Colligative properties: Those properties of solution which depend upon the number of particles of solute in solution.

Equivalent Conductance: It is defined as the conductance of a solution containing from equivalent of an electrolyte, the solution containing from equivalent of an electrolyte, the solution being contained between two plane electrodes/ cm apart.

A gram molecular weight: A substance is its molecular weight expressed in grams.

A gram atomic weight: It is the atomic weight of an element expressed in grams.

An equivalent weight: Equivalent weight of a substance is that weight which is directly or indirectly equivalent in reacting power of an atom of hydrogen.

A milli equivalent weight: It is one thousandth of the equivalent weight.

A gram equivalent weight: It is equivalent weight expressed in grams.

A molar solution: It is one containing gram molecular weight of a substance in 1000 ml of solvent

A normal Solution: It is one that contains 1 g equivalent weight of a substance in 1 litre of solution or a gram milliliter of a solution.

Molality: A molal solution contains one gram molecular weight of a substance dissolved in 1000 g of water

Mesh: The number of apertures per linear inch in an oven or electroformed metal screen or sieve.

Aliquot: A portion of sample which is a specific portion of the total sample.

Chapter 54

Conversion Factors

1.	Eq. Wt.	=	Atomic mass/Electro-valance
2.	M. Eq.	=	Eq. wt./1000
3.	ml x N	=	m. eq.
4.	mg/litre	=	ppm (only water is the solvent)
5.	% x 10000	=	ppm
6.	ppm/10000	=	%
7.	ppm x 2	=	lbs/ac
8.	ppm x 2.24	=	kg/ha
9.	m. eq. x eq. wt.	=	mg
10.	m. eq./litre x eq. wt.	=	ppm
11.	Normality of the solution (for 100 ml)	=	$\frac{\text{V x basicity x \% assay x specific gravity}}{\text{100 x mole. Wt.}}$ Where, V = Volume of the reagent
12.	Normality	=	$\frac{\text{Wt. / liter}}{\text{Eq. Wt.}}$
13.	Wt./litre	=	millimohs/cm
14.	Mohs/cm x 10^6	=	micro mohs/cm
15.	ppm	=	0.64 x ECx 10^6
16.	ppm/eq. wt.	=	m. eq./lit
17.	mg./lit		m. eq./lit
18.	m. eq. to %	=	mg
	m. eq. x eq. wt.	=	mg
	m. eq./100 gm x m. eq. wt.	=	%

19. m. eq. / lit x eq. wt. = ppm

20. Normality = $\frac{\text{Wt. / litre}}{\text{Eq. Wt.}}$

21. Molarity = $\frac{\text{Wt. / litre}}{\text{Mol. Wt.}}$

22. Molality = $\frac{\text{Wt. / litre}}{\text{Mol. Wt.}}$

23. Conversion factors from element to its molecular form :

Per cent P_2O_5 x 0.44 = % P
Per cent P x 2.29 = % P_2O_5
Per cent K_2O x0.83 = %K
Per cent K x 1.20 = %K_2O
Per cent Ca x 1.40 = %CaO
Per cent Mg x 1.67 = %MgO
1 acre = 4840 Sq. Yards. (Or) 4047 Sq. m
43560 acre OR 0.404686
1 ha. = 2.471 acre or 10000 Sq. m
1 lb = 0.4536 Kgs
1 kg = 2.2046 lbs
1 Acre Furrow Slice (AFS) (6" depth) = 2 million pounds (m. lbs)
1 Hectare Furrow Slice (HaFS) = 2 million kgs.

Data on the strength of aqueous solution of the common acids and aqueous NH_3

Reagent	% by weight	Approximate specific gravity	Normality	Volume required to make 1 liter of 1 N solution
HCl (hydrochloric acid)	35.0	1.18	11.3	89.0
HNO_3(nitric acid)	70.0	1.42	16.0	63.0
H_2SO_4(sulphuric acid)	96.0	1.84	36.0	28.0
H_3PO_4(phosphoric acid)	85.0	1.69	41.0	23.0
CH_3COOH(Acitic acid)	88.5	1.05	17.4	58.0
Aq NH_3(Ammonia)	27.0	0.90	14.3	71.0

Conversion Factor for non SI and SI Units

To covert A into B multiply by	A (SI Unit)	B (Non SI Unit)	To convert B in to A multiply by
Length			
30.28	Meter, m	Foot, ft	0.304
0.0394	Millimetre, mm (10^{-3}m)	Inch, in	25.4
1.0	Micrometre, um (10^{-6}m)	Micron, u	1.0
10	Nanometre, nm (10^{-9}m)	Angstrom, A°	0.1
2.10	Metre per second m/sec	Mile per hour	0.447
Area			
20.47	Hectare, ha	acre	0.405
2.47×10^{-4}	Sq. Mtr. $(10^3\ m)^2$	acre	4.05×10^3
10.76	Square mm. $(10^3\ m)^2$	Square foot, ft^2	9.29×10^2
1.55×10^{-3}	Square km. $(10^3\ m)^2$	Square inch, in^2	645
0.39	Square km. $(10^3\ m)^2$	Square mile, mi^2	2.59
Volume			
0.00973	Cubic metre, (m^3)	Acre inch	102.8
35.7	Metre^3, (m^3)	Cubic foot, ft^3	0.028
0.026	Litter, L ($10^{-3}\ m^3$)	Gallon	3.78
0.11	Litter per ha, (L/ha)	Gallon per acre	9.35
1.00	Mega gram mt^{-3} (Mg, m^{-3})	Gram cm^{-3}, g cm^{-3}	1.00
Mass, Yield, or Rate			
10.102	Tonne (metric), t	Ton (U.S), T	0.9077
2.205	Kilo gram, Kg	Pound, lb	0.454
2.2×10^{-3}	Gram, g ($10^{-3)}$	Pound, lb	454
0.035	Gram, g	Ounce (avdp), oz	28.4
0.89	Kg ha^{-1}	lb $acre^{-1}$	1.12
893	Mg ha^{-1}	Pound per acre	1.12×10^{-3}
Pressure			
9.9	Mega pascal, Mpa (10^6 pa)	Atmosphere	0.101
10.0	Mega pascal, Mpa (10^6 pa)	Bar	0.1
2.09×10^{-2}	Pascal, Pa	lb ft^{-2}	47.9
1.45×10^{-4}	Pascal, Pa	lb in^{-2}	6.90×10^3
Temperature			
1.00	Kelvin, k	Celsius, C (C+273)	1.00
(9/5°C)+3	Celsius (-17.8°C)	Fahrenheit. °F	5/9 (F-32)
	0°C	32°F	
	20°C	68°F	
	100°C	212°F	

Energy, Work, Quality of Heat			
10^5	Newton, N	Dyne	10^{-5}
10^7	Joule, J	Erg	10^{-7}
0.74	Joule, J	Foot-pound	1.36
Angle			
571	Radian, (rad)	Degree (angle).	1.75×10^{-2}
Electrical Conductivity			
10	Siemen per meter, Sm^{-1}	m mhos cm^{-1}	0.1
Water measurement			
80.11	Hectare-meter, ha-m	Acre-feet. Acre-ft	0.1233
97.28	Hectare-meter, ha-m	Acre-inches, acre-in	0.01028
0.081	Hectare-cms, ha-cm	Acre-feet, acre-ft	12.33
0.00973	Cubic meter, m^3	Acre-inches, acre-ft	102.8
0.00981	M^3 ha^{-1}	$Feet^3$ sec^{-1}	101.94
4.40	M^3 ha^{-1}	U.S. gallons min^{-1}	0.227
Plant nutrient conversion			
	Elemental	Oxide	
2.29	P	P_2O_5	0.44
1.20	K	K_2O	0.83
1.40	Ca	CaO	0.71
1.66	Mg	MgO	60

Approximate wave length of colours

Colour	Wave length (nm)
1. Violet	440-435
2. Blue	435-480
3. Greenish blue	480-490
4. Bluish green	490-500
5. Green	500-560
6. Yellowish green	560-580
7. Yellow	580-595
8. Orange	595-610
9. Red	610-760

1μ (micron) = 10^{-4}, 1mμ (milli micron) = 10^{-7}, 1A0 (Angstrom) = 10^{-8}cm

Indicators for acid base titration

For titrating	Indicators used
Strong acid against strong base	Phenolphthalein, methyl red or methyl orange
Strong acid against weak base	methyl red or methyl orange
Weak acid against strong base	Phenolphthalein

Appendix

Table 55.1: The depth of sampling for silt + clay (> 0.02) at 5minutes and for clay (< 0.002mm) at 3 hours 30 minutes

Temperature	Silt + Clay	Clay
10	8.2	5.4
11	8.5	5.6
12	8.7	5.7
13	9.0	5.9
14	9.2	6.1
15	9.4	6.2
16	9.7	6.4
17	9.9	6.6
18	10.1	6.7
19	10.4	6.9
20	10.7	7.1
21	11.0	7.2
22	11.2	7.4
23	11.5	7.6
24	11.8	7.8
25	12.1	8.0
26	12.3	8.1
27	12.6	8.3
28	12.9	8.5
29	13.2	8.7
30	13.5	8.9
31	13.8	9.1
32	14.1	9.3
33	14.3	9.5
34	15.7	9.7
35	14.9	9.9

Table 55.2: The conversion table for converting % transmittance to absorbance (OD)

% Transmittance	Absorbance	% Transmittance	Absorbance
100	0.000	50	0.301
99	0.004	49	0.310
98	0.009	48	0.319
97	0.013	47	0.328
96	0.018	46	0.337
95	0.022	45	0.347
94	0.027	44	0.357
93	0.032	43	0.367
92	0.036	42	0.377
91	0.041	41	0.387
90	0.046	40	0.398
89	0.051	39	0.409
88	0.056	38	0.420
87	0.061	37	0.432
86	0.066	36	0.444
85	0.071	35	0.456
84	0.076	34	0.469
83	0.081	33	0.482
82	0.086	32	0.495
81	0.092	31	0.509
80	0.097	30	0.523
79	0.102	29	0.538
78	0.108	28	0.552
77	0.114	27	0.569
76	0.119	26	0.585
75	0.125	25	0.602
74	0.131	24	0.620
73	0.137	23	0.638
72	0.143	22	0.658
71	0.149	21	0.678

70	0.155	20	0.699
69	0.161	19	0.721
68	0.168	18	0.745
67	0.174	17	0.770
66	0.181	16	0.796
65	0.187	15	0.824
64	0.194	14	0.850
63	0.201	13	0.886
62	0.208	12	0.921
61	0.215	11	0.959
60	0.222	10	1.000
59	0.229	9	1.046
58	0.237	8	1.097
57	0.244	7	1.155
56	0.252	6	1.222
55	0.260	5	1.301
54	0.268	4	1.398
53	0.276	3	1.523
52	0.284	2	1.699
51	0.292	1	2.000